Tuncer Cebeci

Convective Heat Transfer Second Edition
Solutions Manual and Computer Programs

Springer-Verlag Berlin Heidelberg GmbH

Tuncer Cebeci

Convective Heat Transfer

Second Revised Edition

Solutions Manual
and
Computer Programs

With 27 Figures, 34 Tables, 122 Problems
and a CD-ROM

HORIZONS
PUBLISHING

Tuncer Cebeci
810 Rancho Drive
Long Beach, CA 90815, USA
TuncerC@aol.com

The first edition (soft cover) of this manual was published in 1989 by Springer-Verlag New York Berlin Heidelberg under the title: Tuncer Cebeci, *Solutions Manual and Computer Programs for Physical and Computational Aspects of Convective Heat Transfer by T. Cebeci and P. Bradshaw.*

Additional material to this book can be downloaded from http://extras.springer.com

Library of Congress Cataloging-in-Publication Data
Cebeci, Tuncer
Convective heat transfer. Solutions manual and computer programs / Tuncer Cebeci. –
2nd rev. ed. p. cm.
ISBN 0-9668461-5-X
1. Heat – Convection – Problems, exercises, etc. 2. Heat–Convection–Computer programs. I. Title
TJ260.C35 2002b   621.402'2–dc21   2002068512

ISBN 978-3-662-06408-5   ISBN 978-3-662-06406-1 (eBook)
DOI 10.1007/978-3-662-06406-1

All rights reserved. This work may not be translated or copied in whole or in part without the written permission of the publisher (Horizons Publishing Inc., 810 Rancho Drive, Long Beach, CA 90815, USA) except for brief excerpts in connection with reviews or scholarly analysis. Use in connection with any form of information storage and retrieval, electronic adaptation, computer software, or by similar or dissimilar methodology now known or hereafter developed is forbidden.

© Springer-Verlag Berlin Heidelberg 2002
Originally published by Springer-Verlag Berlin Heidelberg New York in 2002
Softcover reprint of the hardcover 2nd edition 2002

The use of general descriptive names, trade names, trademarks, etc., in this publication, even if the former are not especially identified, is not to be taken as a sign that such names, as understood by the Trade Marks and Merchandise Marks Act, may accordingly be used freely by anyone.

Please note: All rights pertaining to the Computer Programs are owned exclusively by the author and Horizons Publishing Inc. The publisher and the author accept no legal responsibility for any damage caused by improper use of the programs. Although the programs have been tested with extreme care, errors cannot be excluded.

Typeset in MS Word by the author. Edited and reformatted by Kurt Mattes, Heidelberg, Germany, using LaTeX.

Cover design: Erich Kirchner, Heidelberg, Germany

Printed on acid-free paper   5 4 3 2 1 0

Preface

This book presents the solutions of the problems in my book *Convective Heat Transfer, Second Edition*. The book also contains computer programs to solve homework problems 4.19–4.21, 4.24, 5.4–5.7, 7.5, 7.6, 7.10, 9.1–9.6 and 11.1–11.6 on the CD accompanying the book. Included on the CD are computer programs based on integral methods described in Chapter 11 of the *Convective Heat Transfer* book. These computer programs are for two-dimensional flows and include Thwaites' method for momentum transfer, Smith-Spalding method for heat transfer, both for laminar flows; Michel's method for predicting transition; Head's method for momentum transfer and Ambrok's method for heat transfer, both for turbulent flows and a panel method to predict inviscid flow fields around airfoils.

Many people have contributed to the preparation of this book. Special thanks are due to Dr. J. P. Shao and H. H. Chen who provided me with significant help in preparing the CD and computer programs.

Indian Wells *Tuncer Cebeci*
March 2002

Contents

1. Introduction 1
2. Conservation Equations for Mass, Momentum, and Energy   3
3. Boundary-Layer Equations 11
4. Laminar Boundary Layers 21
5. Laminar Duct Flows 53
6. Turbulent Boundary Layers 71
7. Turbulent Duct Flows 87
8. Buoyant Flows 101
9. Finite-Difference Solution of Boundary-Layer Equations: Internal Flows 109
11. Computer Programs and Their Applications to Momentum and Heat Transfer Problems 115

Appendix 125
   A. Conversion Factors 125
   B. Physical Properties of Gases, Liquids, Liquid Metals, and Metals 128
   C. Gamma, Beta and Incomplete Beta Functions 141

1
Introduction

1.1 PROBLEM

Calculate the Prandtl number of fluids with the following properties
(a) a liquid metal with $\rho = 879\,\text{kg m}^{-3}$, $\mu = 3.3 \times 10^{-4}\,\text{kg m}^{-1}\text{s}^{-1}$, $\kappa = 66.5 \times 10^{-6}\,\text{m}^2\text{s}^{-1}$;
(b) a gas with $c_p = 0.2415\,\text{Btu lb}^{-1}\,°\text{F}^{-1}$, $\mu = 5.193 \times 10^{-2}\,\text{lb ft}^{-1}\,\text{h}^{-1}$, $k = 1.806 \times 10^{-2}\,\text{Btu h}^{-1}\,\text{ft}^{-1}\,°\text{F}^{-1}$;
(c) a liquid with $\rho = 900\,\text{kg m}^{-3}$, $c_p = 1.902\,\text{kJ kg}^{-1}\,°\text{K}^{-1}$; $\nu = 9 \times 10^{-4}\,\text{m}^2\,\text{s}^{-1}$, $k = 0.143\,\text{W m}^{-1}\,°\text{K}^{-1}$.

SOLUTION

a. With $\rho = 879\,\text{kg/m}^3$,
$\mu = 3.3 \times 10^{-4}\,\text{kg/m s}$,
$\kappa = 66.5 \times 10^{-6}\,\text{m}^2/\text{s}$,
$\text{Pr} = \mu/\rho\kappa = 3.3 \times 10^{-4}/(879.0 \times 66.5 \times 10^{-6}) = 0.0056$.
b. With $c_p = 0.2415\,\text{Btu/lb}\,°\text{F}$,
$\mu = 5.193 \times 10^{-2}\,\text{lb/ft h}$,
$k = 1.806 \times 10^{-2}\,\text{Btu/h}\,°\text{F}$,
$\text{Pr} = \mu c_p/k = 5.193 \times 10^{-2} \times 0.2415/(1.806 \times 10^{-2}) = 0.694$.
c. With $\rho = 900\,\text{kg/m}^3$,
$c_p = 1902\,\text{kJ/kg K}$,
$\nu = 9 \times 10^{-4}\,\text{m}^2/\text{s}$,
$k = 0.143\,\text{W/m K}$,
$\text{Pr} = \mu c_p/k = \rho\nu c_p/k = 900.0\times(9\times 10^{-4})\times 1.902\times 10^3/0.143 = 10773.6$.

1.2 PROBLEM

Calculate the Reynolds number for the following cases: (a) air with a freestream velocity of $100\,\text{ft s}^{-1}$, a pressure of $14.7\,\text{lb in}^{-2}$, and a temperature of $150\,°\text{F}$ flowing over a 6-inch long flat plate; (b) water with an average velocity of $12\,\mathrm{m\,s^{-1}}$ and an average temperature of $25\,^\circ\mathrm{C}$ flowing in a circular duct of $0.25\,\mathrm{m}$ in diameter; (c) glycerin at $30\,^\circ\mathrm{C}$ with a velocity of $10\,\mathrm{m\,s^{-1}}$ at a distance of $1\,\mathrm{m}$ from the leading edge of a flat plate.

SOLUTION

a. With $u_\infty = 100\,\mathrm{ft/s}$, $L = 0.5\,\mathrm{ft}$ and $\nu = 2.15 \times 10^{-4}\,\mathrm{ft^2/s}$
   for air at $T = 150\,^\circ\mathrm{F}$ and $p = 14.7\,\mathrm{lb/in^2}$,
   $R_L = u_\infty L/\nu = 100 \times 0.5/(2.15 \times 10^{-4}) = 2.33 \times 10^5$.
b. With $\nu = 0.925 \times 10^{-6}\,\mathrm{m^2/s}$ for water at $T = 25\,^\circ\mathrm{C}$,
   $p = 14.7\,\mathrm{lb/in^2}$, $u_m = 12\,\mathrm{m/s}$, $d = 0.25\,\mathrm{m}$,
   $R_d = u_m d/\nu = 12.0 \times 0.25/(0.925 \times 10^{-6}) = 3.24 \times 10^6$.
c. With $u_e = 10\,\mathrm{m/s}$, $x = 1\,\mathrm{m}$ and $\nu = 5.0 \times 10^{-4}\,\mathrm{m^2/s}$ for glycerine
   at $T = 3\,^\circ\mathrm{C}$, $R_x = u_e x/\nu = 10.0 \times 1.0/(5.0 \times 10^{-4}) = 2.0 \times 10^4$.

1.3 PROBLEM

In Problem 1.2(a), if the local skin-friction coefficient c_f is given by the formula,

$$c_f = \frac{0.664}{\sqrt{R_x}}$$

where $R_x = u_e x/\nu$, find the (a) wall shear, τ_w, (b) wall heat flux, $\dot{q}_w$. Take $T_w - T_e = 10\,^\circ\mathrm{F}$.

SOLUTION

a. With $u_e = 100\,\mathrm{ft/s}$ and $\varrho = 0.065\,\mathrm{lbm/ft^3}$
   for air at $T = 150\,^\circ\mathrm{F}$ and $p = 14.7\,\mathrm{lb/in^2}$,
   $\tau_w = \varrho u_e^2 0.332 (R_x)^{-1/2} = 0.065 \times 100^2 \times 0.322/\sqrt{2.33 \times 10^5}$
   $= 0.447\,\mathrm{lbm/ft\,s^2}$
b. Since Pr for air is nearly unity, Reynolds analogy should be a good approximation. Then form

$$\mathrm{St}/(c_f/2) = 1.0 = \dot{q}_w/[c_f/2 \varrho c_p (T_w - T_e) u_e],$$

$\dot{q}_w = 0.322/\sqrt{2.33 \times 10^5} \times 0.065 \times 0.2410 \times 10 \times 100 = 0.0108\,\mathrm{Btu/ft^2\,s}$.

2. Conservation Equations for Mass, Momentum, and Energy

2.1 Problem

(a) Extend the control volume analysis used to derive the continuity equation to three-dimensional unsteady flows and show that Eq. (2.2.1) can be written as

$$\frac{\partial \varrho}{\partial t} + \frac{\partial}{\partial x}(\varrho u) + \frac{\partial}{\partial y}(\varrho v) + \frac{\partial}{\partial z}(\varrho w) = 0. \qquad (\text{P2.1})$$

(b) Deduce from Eq. (P2.1) the mass conservation equations for three-dimensional steady or unsteady constant-density flow and for three-dimensional steady, variable-density flows corresponding to Eqs. (2.2.2) and (2.2.3) for two-dimensional flows.

(c) Show that Eq. (P2.1) can be rewritten as a transport equation for ϱ, using the three-dimensional version of the d/dt operator defined in Eq. (2.3.9).

Solution

a. The derivation of the equation is straightforward; there is no need to elaborate on the details of the algebra.

b. Eq. (P2.1) may be written as

$$\frac{\partial \varrho}{\partial t} + u\frac{\partial \varrho}{\partial x} + v\frac{\partial \varrho}{\partial y} + w\frac{\partial \varrho}{\partial z} + \varrho\left(\frac{\partial u}{\partial x} + \frac{\partial v}{\partial y} + \frac{\partial w}{\partial z}\right) = 0. \qquad (1)$$

For constant-density flow

$$\frac{\partial \varrho}{\partial t} = \frac{\partial \varrho}{\partial x} = \frac{\partial \varrho}{\partial y} = \frac{\partial \varrho}{\partial z} = 0.$$

Hence

$$\frac{\partial u}{\partial x} + \frac{\partial v}{\partial y} + \frac{\partial w}{\partial z} = 0. \qquad (2)$$

For steady but variable-density flows, $\partial\varrho/\partial t = 0$, and (P2.1) becomes

$$\frac{\partial}{\partial x}(\varrho u) + \frac{\partial}{\partial y}(\varrho v) + \frac{\partial}{\partial z}(\varrho w) = 0. \qquad (3)$$

c. Since by definition

$$\frac{d\varrho}{dt} \equiv \frac{\partial\varrho}{\partial t} + u\frac{\partial\varrho}{\partial x} + v\frac{\partial\varrho}{\partial y} + w\frac{\partial\varrho}{\partial z},$$

Eq. (1) may be written as

$$\frac{d\varrho}{dt} = \frac{\partial\varrho}{\partial t} + u\frac{\partial\varrho}{\partial x} + v\frac{\partial\varrho}{\partial y} + w\frac{\partial\varrho}{\partial z} = -\varrho\left(\frac{\partial\varrho}{\partial x} + \frac{\partial\varrho}{\partial y} + \frac{\partial\varrho}{\partial z}\right)$$

or as

$$\frac{d\varrho}{dt} + \varrho\nabla\cdot\boldsymbol{v} = 0,$$

which is, by definition, the transport equation for ϱ.

2.2 Problem

By repeating the arguments used to derive the x-component momentum equation, Eq. (2.3.10), show that the y-component momentum equation is as given in Eq. (2.3.11). Check your answer by "rotating" the coordinates in the x-component momentum equation (i.e. changing x to y and y to x, u to v and v to u, througout).

Solution

Consider Eq. (2.3.1) and write it for the y-component of the momentum equation, that is,

RATE OF INCREASE OF y-COMPONENT MOMENTUM OF FLUID IN CV		RATE OF FLOW OF y-COMPONENT MOMENTUM INTO CV		RATE OF FLOW OF y-COMPONENT MOMENTUM OUT OF CV		SUM OF y-COMPONENTS OF FORCES APPLIED TO FLUID IN CV
	−		+		=	

(1)

Then write Eqs. (2.3.2), (2.3.3) and (2.3.6) as, respectively,

$$\boxed{\text{RATE OF INCREASE OF } y\text{-COMPONENT MOMENTUM OF FLUID IN CV}} = \frac{\partial}{\partial t}(\varrho v)dx\,dy\,dz. \qquad (2)$$

2. Conservation Equations for Mass, Momentum, and Energy

$$\begin{bmatrix} \text{RATE OF FLOW} \\ \text{OF } y\text{-COMPONENT} \\ \text{MOMENTUM OUT} \\ \text{OF CV} \end{bmatrix} - \begin{bmatrix} \text{RATE OF FLOW} \\ \text{OF } y\text{-COMPONENT} \\ \text{MOMENTUM} \\ \text{INTO CV} \end{bmatrix} = \left[\frac{\partial}{\partial x}(\varrho u v) + \frac{\partial}{\partial y}(\varrho v^2) \right] dx\, dy\, dz.$$

(3)

$$\begin{bmatrix} \text{SUM OF } y \text{ COMPONENTS} \\ \text{OF FORCES APPLIED} \\ \text{TO FLUID IN CV} \end{bmatrix} = \left(-\frac{\partial p}{\partial y} + \frac{\partial \sigma_{yx}}{\partial x} + \frac{\partial \sigma_{yy}}{\partial y} + \varrho f_y \right) dx\, dy\, dz. \quad (4)$$

Writing Eqs. (2)–(4) according to Eq. (1) and dividing both sides by $dx\, dy\, dz$, we get

$$\frac{\partial}{\partial t}(\varrho v) + \frac{\partial}{\partial x}(uv) + \frac{\partial v^2}{\partial y} = -\frac{\partial p}{\partial y} + \frac{\partial \sigma_{yx}}{\partial x} + \frac{\partial \sigma_{yy}}{\partial y} + \varrho f_y. \tag{5}$$

Multiply the continuity equation by v and add it to Eq. (5). After simplifying and dividing both sides by ϱ, we get

$$\frac{\partial v}{\partial t} + u\frac{\partial v}{\partial x} + v\frac{\partial v}{\partial y} = -\frac{1}{\varrho}\frac{\partial p}{\partial y} + \frac{1}{\varrho}\left(\frac{\partial \sigma_{yx}}{\partial x} + \frac{\partial \sigma_{yy}}{\partial y} \right) + f_y$$

or

$$\frac{dv}{dt} = -\frac{1}{\varrho}\frac{\partial p}{\partial y} + \frac{1}{\varrho}\left(\frac{\partial \sigma_{yx}}{\partial x} + \frac{\partial \sigma_{yy}}{\partial y} \right) + f_y. \tag{6}$$

To check the answer by "rotating" the coordinates in the x-momentum equation consider Eq. (2.3.10)

$$\frac{du}{dt} = -\frac{1}{\varrho}\frac{\partial p}{\partial x} + \frac{1}{\varrho}\left(\frac{\partial \sigma_{xx}}{\partial x} + \frac{\partial \sigma_{xy}}{\partial y} \right) + f_x. \tag{2.3.10}$$

Rotating the coordinates, we get Eq. (6).

2.3 PROBLEM

Show that

$$\frac{\partial}{\partial t} + u\frac{\partial}{\partial x} + v\frac{\partial}{\partial y} \equiv \frac{d}{dt}$$

represents the rate of change, with respect to time, as seen by an observer following the motion of a fluid element.

SOLUTION

Suppose that in a short time dt, a fluid element moves from p to p', as shown in the figure below. Then a quantity – such as temperature, say – that has the value ϕ at location p and time t will have the value

$$\phi + \frac{\partial \phi}{\partial x}dx + \frac{\partial \phi}{\partial y}dy$$

at point p' and time t, and the value

$$\phi + \frac{\partial \phi}{\partial t} dt + \frac{\partial \phi}{\partial x} dx + \frac{\partial \phi}{\partial y} dy \equiv \phi',$$

say at point p' and time $t + dt$ (where $\partial \phi / \partial t$ is the time derivative at a fixed point, strictly at p'). Since $dx = u\,dt$, $dy = v\,dt$ we can write

$$\phi' = \phi + \left(\frac{\partial \phi}{\partial t} + \frac{\partial \phi}{\partial x} + \frac{\partial \phi}{\partial y} \right) dt$$

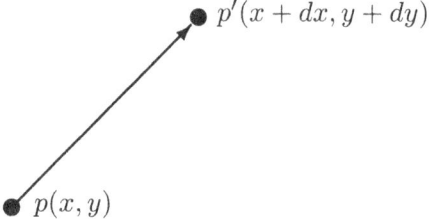

2.4 PROBLEM

By extending the arguments used to derive the x-component momentum equation for two-dimensional flow, Eq. (2.3.8), show that in three-dimensional flow the equation becomes

$$\frac{\partial u}{\partial t} + u \frac{\partial u}{\partial x} + v \frac{\partial u}{\partial y} + w \frac{\partial u}{\partial z} = -\frac{1}{\varrho} \frac{\partial p}{\partial x} + \frac{1}{\varrho} \left(\frac{\partial \sigma_{xx}}{\partial x} + \frac{\partial \sigma_{xy}}{\partial y} + \frac{\partial \sigma_{xz}}{\partial z} \right) + f_x$$

where in a Newtonian fluid

$$\sigma_{xz} = \frac{\partial u}{\partial z} + \frac{\partial w}{\partial x}.$$

SOLUTION

For a three-dimensional flow,

$$\boxed{\begin{array}{l}\text{RATE OF FLOW}\\ \text{OF } x\text{-COMPONENT}\\ \text{MOMENTUM}\\ \text{INTO CV}\end{array}} = \varrho u^2\, dy\, dz + \varrho uv\, dx\, dz + \varrho uw\, dx\, dy$$

$$\boxed{\begin{array}{l}\text{RATE OF FLOW}\\ \text{OF } x\text{-COMPONENT}\\ \text{MOMENTUM OUT}\\ \text{OF CV}\end{array}} = \left[\varrho u^2 + \frac{\partial}{\partial x}(\varrho u^2)\, dx \right] dy\, dz$$

$$+ \left[\varrho uv + \frac{\partial}{\partial y}(\varrho uv)\, dy \right] dx\, dz + \left[\varrho uw + \frac{\partial}{\partial z}(\varrho uw)\, dz \right] dy\, dz.$$

Therefore,

| RATE OF FLOW OF x-COMPONENT MOMENTUM OUT OF CV | $-$ | RATE OF FLOW OF x-COMPONENT MOMENTUM INTO CV |

$$= \left[\frac{\partial}{\partial x}(\varrho u^2) + \frac{\partial}{\partial y}(\varrho uv) + \frac{\partial}{\partial z}(\varrho uw)\right] dx\, dy\, dz$$

| SUM OF x COMPONENTS OF FORCES APPLIED TO FLUID IN CV | $= \left(-\dfrac{\partial p}{\partial x} + \dfrac{\partial \sigma_{xx}}{\partial x} + \dfrac{\partial \sigma_{xy}}{\partial y} + \dfrac{\partial \sigma_{xz}}{\partial z} + \varrho f_x\right) dx\, dy\, dz.$ |

Therefore, the x-component of the momentum equation for a three-dimensional flow is

$$\frac{\partial \varrho u}{\partial t} + \frac{\partial}{\partial x}(\varrho u^2) + \frac{\partial}{\partial y}(\varrho uv) + \frac{\partial}{\partial x}(\varrho uw) = -\frac{\partial p}{\partial x} + \frac{\partial \sigma_{xx}}{\partial x} + \frac{\partial \sigma_{xy}}{\partial y} + \frac{\partial \sigma_{xz}}{\partial z} + \varrho f_x. \tag{1}$$

Multiplying the continuity equation by u,

$$\Rightarrow \quad u\frac{\partial \varrho}{\partial t} + u\frac{\partial \varrho u}{\partial x} + u\frac{\partial \varrho v}{\partial y} + u\frac{\partial \varrho w}{\partial z} = 0. \tag{2}$$

Adding Eqs. (1) and (2), we get

$$\varrho\frac{\partial u}{\partial t} + \varrho u\frac{\partial u}{\partial x} + \varrho v\frac{\partial u}{\partial y} + \varrho w\frac{\partial u}{\partial z} = -\frac{\partial p}{\partial x} + \frac{\partial \sigma_{xx}}{\partial x} + \frac{\partial \sigma_{xy}}{\partial y} + \frac{\partial \sigma_{xz}}{\partial z} + \varrho f_x.$$

Dividing by ϱ

$$\frac{\partial u}{\partial t} + u\frac{\partial u}{\partial x} + v\frac{\partial u}{\partial y} + w\frac{\partial u}{\partial z} = -\frac{1}{\varrho}\frac{\partial p}{\partial x} + \frac{1}{p}\left(\frac{\partial \sigma_{xx}}{\partial x} + \frac{\partial \sigma_{xy}}{\partial y} + \frac{\partial \sigma_{xz}}{\partial z}\right) + f_x.$$

2.5 Problem

Show that the term $\partial(\sigma_{ij}u_i)/\partial x_j$ in Eq. (2.4.13) can be expanded, in two-dimensional flow, to

$$\frac{\partial \sigma_{xx} u}{\partial x} + \frac{\partial \sigma_{yx} v}{\partial x} + \frac{\partial \sigma_{xy} u}{\partial y} + \frac{\partial \sigma_{yy} v}{\partial y}$$

SOLUTION

$$\frac{\partial}{\partial x_j}\delta_{ij}u_i = \frac{\partial}{\partial x_1}\delta_{11}u_1 + \frac{\partial}{\partial x_2}\delta_{12}u_1 + \frac{\partial}{\partial x_1}\delta_{21}u_2 + \frac{\partial}{\partial x_2}\delta_{22}u_2$$

$$x_1 \to x \quad u_1 \to u \quad \delta_{11} \to \delta_{xx} \quad \delta_{21} \to \delta_{yx}$$
$$x_2 \to y \quad u_2 \to v \quad \delta_{12} \to \delta_{xy} \quad \delta_{22} \to \delta_{yy}$$

$$\frac{\partial}{\partial x_j}\delta_{ij}u_i = \frac{\partial}{\partial x}(\delta_{xx}u) + \frac{\partial}{\partial y}(\delta_{xy}u) + \frac{\partial}{\partial x}(\delta_{yx}v) + \frac{\partial}{\partial y}(\delta_{yy}v)$$

2.6 PROBLEM

If we multiply the x-component of the momentum equation Eq. (2.3.8) by u, and use the definition of the substantial-derivative symbol d/dt given by Eq. (2.3.9), we get

$$\frac{d}{dt}\left(\frac{1}{2}u^2\right) = -\frac{u}{\varrho}\frac{\partial p}{\partial x} + \frac{u}{\varrho}\left[\frac{\partial \sigma_{xx}}{\partial x} + \frac{\partial \sigma_{xy}}{\partial y}\right] + uf_x. \tag{P2.2}$$

Adding the corresponding equation for $d/dt(1/2v^2)$, which can be most easily derived by changing the variables in Eq. (P2.2) in cylic order, show that the resulting expression, with V denoting the resulting velocity, can be written as

$$\frac{d}{dt}\left(\frac{1}{2}V^2\right) = -\frac{1}{\varrho}\left(u\frac{\partial p}{\partial x} + v\frac{\partial p}{\partial y}\right) + \frac{u}{\varrho}\left(\frac{\partial \sigma_{xx}}{\partial x} + \frac{\partial \sigma_{xy}}{\partial y}\right) + \frac{v}{\varrho}\left(\frac{\partial \sigma_{yx}}{\partial x} + \frac{\partial \sigma_{yy}}{\partial y}\right)$$
$$+ uf_x + vf_y. \tag{P2.3}$$

Equation (P2.3) is known as the kinetic energy equation. Its left-hand side represents the rate of increase of kinetic energy per unit mass of the fluid, as the fluid moves along a streamline. The terms on the right-hand side which can be written in tensor notation as

$$\frac{u_i}{\varrho}\left(-\frac{\partial p}{\partial x_i} + \frac{\partial \sigma_{ij}}{\partial x_j} + \varrho f_i\right)$$

represent, respectively, the rates at which work is done on unit mass of the fluid by the pressure, by the viscous or turbulent σ-stresses, and by the body force per unit mass, f.

SOLUTION

Use the definition of the substantial derivative, Eq. (2.3.9), multiply Eq. (2.3.8) with u, Eq. (2.3.11) with v, and add the resulting expressions to obtain the desired equation.

2.7 PROBLEM

By adding $d(p/\varrho)/dt$ to the total-energy equation for $e + 1/2(u^2 + v^2)$, derive an equation for the total enthalpy $H \equiv h + 1/2(u^2 + v^2)$.

Hint: Follow the derivation of the *static*-enthalpy equation from the *internal*-energy equation, and recall that d/dt is the transport operator defined in Eq. (2.3.9).

SOLUTION

Note the total-energy equation for two-dimensional flow is given by (2.4.12). The term, $d(p/\varrho)/dt$ can be expanded as

$$\frac{d(p/\varrho)}{dt} = \frac{1}{\varrho}\frac{dp}{dt} - \frac{p}{\varrho^2}\frac{d\varrho}{dt} = \frac{1}{\varrho}\frac{dp}{dt} + \frac{p}{\varrho}\left(\frac{\partial u}{\partial x} + \frac{\partial v}{\partial y}\right)$$

$$= \frac{1}{\varrho}\left(u\frac{\partial p}{\partial x} + v\frac{\partial p}{\partial y}\right) + \frac{p}{\varrho}\left(\frac{\partial u}{\partial x} + \frac{\partial v}{\partial y}\right)$$

$$= \frac{1}{\varrho}\left[\frac{\partial}{\partial x}(pu) + \frac{\partial}{\partial y}(pv)\right]$$

and is equal to the second term on the right-hand side of (2.4.12). Thus the total enthalpy equation can be written as

$$\frac{d}{dt}\left[e + \frac{p}{\varrho} + \frac{1}{2}(u^2 + v^2)\right] \equiv \frac{dH}{dt} =$$

$$= -\frac{1}{\varrho}\left(\frac{\partial \dot{q}_x}{\partial x} + \frac{\partial \dot{q}_y}{\partial y}\right) + \frac{1}{\varrho}\frac{\partial}{\partial x_j}(\sigma_{ij}u_i) + uf_x + vf_y.$$

Dealing with total enthalpy instead of total energy has the same advantage as dealing with (static) enthalpy rather than internal energy: the pressure does not appear explicitly.

2.8 PROBLEM

Find the mean parts u and T, the fluctuating (time-dependent) parts u' and T', the mean-square fluctuations $\overline{u'^2}$ and $\overline{T'^2}$, and the velocity-temperature mean product ("covariance") $\overline{u'T'}$ for the following variations of instantaneous velocity and temperature with time.

(a) $u + u' = a + b\sin\omega t$, $T + T' = c + d\sin(\omega t - \phi)$
(b) $u + u' = a + b\sin^2\omega t$, $T + T' = c + d\sin^2(\omega t - \phi)$.

Note that the time average can be taken over just one cycle, $0 < \omega t < 2\pi$.

SOLUTION

a. $u = a$, $u' = b\sin\omega t$, $T = c$, $T' = d\sin(\omega t - \phi)$, $\overline{u'^2} = b^2/2$, $\overline{T'^2} = d^2/2$, $\overline{u'T'} = bd/2\cos\phi$ (expand $\sin(\omega t - \phi) = \sin\omega t\cos\phi - \sin\phi\cos\omega t$ and note that the average of $\sin\omega t\cos\omega t$ is zero.)

b. $u = a+b/2$, $u' = b(\sin^2 \omega t - 1/2)$, $T = c+d/2$, $T' = d[\sin^2(\omega t - \phi) - 1/2]$, $\overline{u'^2} = b^2/8$, $\overline{T'^2} = d^2/8$, $\overline{u'T'} = (db/8)\cos\phi$ (note that the average of $\sin^2 \omega t$ is $1/2$).

2.9 PROBLEM

By repeating the arguments used to derive the time-average x-component momentum equation for turbulent flow, Eq. (2.5.7), show that the time-average y-component equation is

$$u\frac{\partial v}{\partial x} + v\frac{\partial v}{\partial y} = -\frac{1}{\varrho}\frac{\partial p}{\partial y} + \frac{1}{\varrho}\left(\frac{\partial \sigma_{xy}}{\partial x} + \frac{\partial \sigma_{yy}}{\partial y}\right) - \frac{1}{\varrho}\left[\frac{\partial}{\partial x}(\varrho\overline{u'v'}) + \frac{\partial}{\partial y}\varrho\overline{v'^2}\right]$$

where terms in the density fluctuation ϱ have been neglected and where the σ terms represent the viscous contributions only.

SOLUTION

Analogous to Eq. (2.5.8), we write the y component of the momentum equation as

$$(\varrho u + \overline{\varrho'u'})\frac{\partial v}{\partial x} + (\varrho u + \overline{\varrho'u'})\frac{\partial v}{\partial y} = -\frac{\partial p}{\partial y} + \frac{\partial \sigma_{yx}}{\partial x} + \frac{\partial \sigma_{yy}}{\partial y} + \varrho\overline{f}_y + \overline{\varrho'f'_y}$$
$$- \left[\frac{\partial}{\partial x}(\varrho\overline{u'v'} + \overline{\varrho'u'v'} + u\overline{\varrho'v'})\right]$$
$$- \left[\frac{\partial}{\partial y}(\varrho\overline{v'^2} + \overline{\varrho'v'^2} + v\overline{\varrho'v'})\right].$$

For constant density all the terms containing ϱ' drop out and the resulting equation becomes the desired equation after both sides are divided by ϱ.

3

Boundary-Layer Equations

3.1 Problem

Show that in a two-dimensional boundary layer whose thickness δ is growing at the rate $d\delta/dx$, the ratio of $\partial^2 u/\partial x^2$ to $\partial^2 u/\partial y^2$, assumed small in the derivation of Eq. (3.2.9), is actually of order $(d\delta/dx)^2$.

Solution

An average value of $\partial u/\partial y$ for a boundary-layer thickness is u_e/δ (u rises from zero at $y=0$ to u_e at $y=\delta$); an average value of $\partial^2 u/\partial y^2$ can be derived by a similar argument: $\partial u/\partial y$ changes from a value of order u_e/δ at $y=0$ to zero at $y=\delta$, so an average value of $\partial(\partial u/\partial y)/\partial y \equiv \partial^2 u/\partial y^2$ is of order u_e/δ^2. The velocity at a fixed distance y from the surface, where y is smaller than the value of δ at the streamwise position considered, falls from u_e at the boundary-layer origin, $x=0$, to a value rather smaller than u_e at the streamwise position considered. Therefore, a typical value of $\partial u/\partial x$ is of order u_e/x or, better, $(u_e/\delta)d\delta/dx$ since if the boundary layer really grew linearly at the rate of $d\delta/dx$, its origin would be at a distance $\delta/(d\delta/dx)$ upstream. As before, the argument can be repeated to give a typical value of $\partial^2 u/\partial x^2$ as of order $(u_e/\delta^2)(d\delta/dx)^2$, so that finally $\frac{\partial^2 u/\partial x^2}{\partial^2 u/\partial y^2}$ is of order $(d\delta/dx)^2$.

3.2 Problem

Show that the thin-shear-layer (TSL) equation for x-component momentum can be rewritten, correct to the TSL approximation, as

$$\left(\frac{\partial P}{\partial s}\right)_\psi = \frac{\partial \sigma_{xy}}{\partial y}$$

where P is the total pressure $P = p + 1/2\varrho(u^2 + v^2)$ and $(\partial/\partial s)_\psi$ is the gradient along a mean streamline (arc distance s). Discuss the order of magnitude of neglected terms.

SOLUTION

$$V\left(\frac{\partial p}{\partial s}\right)_\psi = \left(u\frac{\partial}{\partial x} + v\frac{\partial}{\partial y}\right)P$$

where V ist the resultant velocity,

$$V = (u^2 + v^2)^{1/2}$$

therefore,

$$V\left(\frac{\partial p}{\partial s}\right)_\psi = u\left(\frac{\partial p}{\partial x} + \varrho u\frac{\partial u}{\partial x} + \varrho v\frac{\partial u}{\partial y}\right) + v\left(\frac{\partial p}{\partial y} + \varrho u\frac{\partial v}{\partial x} + \varrho v\frac{\partial v}{\partial y}\right)$$

$$= u\left(\frac{\partial \sigma_{xx}}{\partial x} + \frac{\partial \sigma_{xy}}{\partial y}\right) + v\left(\frac{\partial \sigma_{xy}}{\partial x} + \frac{\partial \sigma_{yy}}{\partial y}\right)$$

or

$$\left(\frac{\partial p}{\partial s}\right)_\psi = \frac{u}{V}\left(\frac{\partial \sigma_{xx}}{\partial x} + \frac{\partial \sigma_{xy}}{\partial y}\right) + \frac{v}{V}\left(\frac{\partial \sigma_{xy}}{\partial x} + \frac{\partial \sigma_{yy}}{\partial y}\right).$$

Now

$$\frac{u}{V} = \frac{u}{(u^2+v^2)^{1/2}} = \left(1 + \frac{1}{2}\frac{v^2}{u^2} + \ldots\right) = \left[1 + O\left(\frac{\delta}{x}\right)^2\right]$$

$$\frac{\partial \sigma_{xx}}{\partial x} = \frac{\sigma_{xx}}{\sigma_{yy}}\frac{\partial \sigma_{xy}}{\partial y}O\left(\frac{\delta}{x}\right), \qquad \frac{v}{V} = \frac{v}{u}\left[1 + O\left(\frac{\delta}{x}\right)^2\right] = O\left(\frac{\delta}{x}\right),$$

$$\frac{\partial \sigma_{xy}}{\partial x} = \frac{\partial \sigma_{xy}}{\partial y}O\left(\frac{\delta}{x}\right), \qquad \frac{\partial \sigma_{yy}}{\partial y} = O\left(\frac{\sigma_{yy}}{\sigma_{xy}}\right)\frac{\partial \sigma_{xy}}{\partial y}.$$

The error in replacing $u/V\,(\partial\sigma_{yy}/\partial y)$ by $(\partial\sigma_{xy}/\partial y)$ is $O(\delta/x)^2$ while the terms in σ_{xx} and σ_{yy} are at most of order δ/x times the main terms (if all σ stresses are of the same order, as in turbulent flow: in laminar flow $\sigma_{xx}, \sigma_{yy} \ll \sigma_{xy}$ and all neglected terms are $O(\delta/x)^2$).

3.3 PROBLEM

Consider a two-dimensional flow in the xy-plane and by using the balance of mass, momentum and energy into a control volume of dimensions dx, dy within the boundary layer, where

$$\frac{\partial^2 u}{\partial y^2} \gg \frac{\partial^2 u}{\partial x^2}, \quad \frac{\partial^2 T}{\partial y^2} \gg \frac{\partial^2 T}{\partial x^2},$$

derive Eqs. (3.2.1), (3.2.10b) and (3.2.11). Show all your assumptions.

3. Boundary-Layer Equations

SOLUTION

Conservation of mass incompressible, steady flow with no sources, sinks within the control volume is

$$(\text{mass flow in}) - (\text{mass flow out}) = 0$$

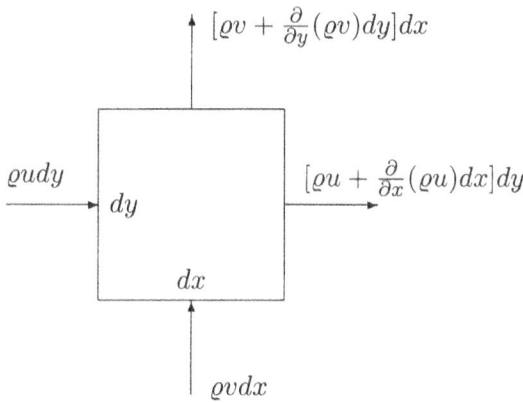

$$\varrho u dy + \varrho v dx - \left[\varrho u + \frac{\partial}{\partial x}(\varrho u)dx\right]dx - \left[\varrho v + \frac{\partial}{\partial y}(\varrho v)dy\right]dx = 0$$

or

$$\frac{\partial u}{\partial x} + \frac{\partial v}{\partial y} = 0$$

x-momentum equation: conservation of momentum.

Rate of change of momentum in x-direction $= \sum(F_x)_{\text{net}}$

$$\sum(F_x)_{\text{net}} = (F_{\text{pressure}} + F_{\text{viscous}} + F_{\text{body}})_{x\text{-direction}}$$

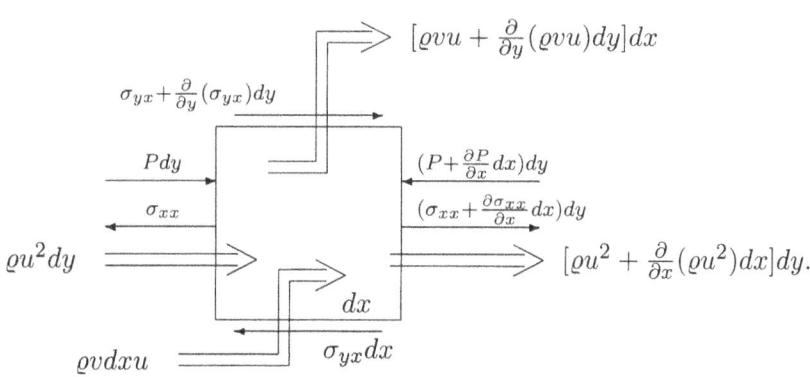

Rate of change of momentum in x-direction

$$\left[\varrho u^2 + \frac{\partial}{\partial x}(\varrho u^2)dx\right]dy + \left[\varrho vu + \frac{\partial}{\partial y}(\varrho vu)dy\right]dx - \varrho u^2 dy - \varrho vu dx$$
$$= \frac{\partial}{\partial x}(\varrho u^2)dxdy + \frac{\partial}{\partial y}(\varrho vu)dydx \qquad (1)$$

$$\sum F_x = -\frac{\partial P}{\partial x}dxdy + \frac{\partial}{\partial x}(\sigma_{xx})dxdy + \frac{\partial}{\partial y}(\sigma_{yx})dydx + \varrho f_x. \qquad (2)$$

Equating (1) and (2)

$$\frac{\partial}{\partial x}(\varrho u^2) + \frac{\partial}{\partial y}(\varrho vu) = -\frac{\partial P}{\partial x} + \frac{\partial}{\partial x}(\sigma_{xx}) + \frac{\partial}{\partial y}(\sigma_{yx}) + \varrho f_x \qquad (3)$$

with

$$\sigma_{xx} = 2\mu\frac{\partial u}{\partial x}, \quad \sigma_{yx} = u\left(\frac{\partial u}{\partial y} + \frac{\partial v}{\partial x}\right),$$

Eq. (3) becomes

$$2\varrho u\frac{\partial u}{\partial x} + \varrho\left(v\frac{\partial u}{\partial y} + u\frac{\partial v}{\partial y}\right) = -\frac{\partial P}{\partial x} + 2\mu\frac{\partial^2 u}{\partial x^2} + \mu\frac{\partial}{\partial y}\left(\frac{\partial u}{\partial y} + \frac{\partial v}{\partial x}\right) + \varrho f_x$$

$$\varrho u\frac{\partial u}{\partial x} + \varrho v\frac{\partial u}{\partial y} + \varrho u\left(\frac{\partial u}{\partial x} + \frac{\partial v}{\partial y}\right) = -\frac{\partial P}{\partial x} + \mu\frac{\partial^2 u}{\partial x^2} + \mu\frac{\partial}{\partial x}\left(\frac{\partial u}{\partial x} + \frac{\partial v}{\partial y}\right)$$
$$+ \mu\frac{\partial^2 u}{\partial y^2} + \varrho f_x. \qquad (4)$$

Using the continuity equation and noting that

$$\frac{\partial^2 u}{\partial x^2} \ll \frac{\partial^2 u}{\partial y^2},$$

and neglecting f_x, Eq. (4) becomes

$$u\frac{\partial u}{\partial x} + v\frac{\partial u}{\partial y} = -\frac{1}{\varrho}\frac{\partial P}{\partial x} - \nu\left(\frac{\partial^2 u}{\partial y^2}\right).$$

Energy equation: conservation of energy

$$E_{\text{in}} + E_{\text{out}} = 0$$

3. Boundary-Layer Equations

$[-k(\frac{\partial T}{\partial y}) + \frac{\partial}{\partial y}(-k\frac{\partial T}{\partial y})dy]dx$ ↑ ↑ $[\varrho v C_p T + \frac{\partial}{\partial y}(\varrho v C_p T)dy]dx$

$(\varrho u C_p T)dy$ ⟶ ⟶ $[\varrho u C_p T + \frac{\partial}{\partial x}(\varrho u C_p T)dx]dy$

$-k\frac{\partial T}{\partial x}dy$ ⟶ ⟶ $[-k(\frac{\partial T}{\partial x}) + \frac{\partial}{\partial x}(-k\frac{\partial T}{\partial x})dx]dy$

$-k\frac{\partial T}{\partial y}dx$ ↑ ↑ $\varrho v C_p T dx$

$$-\frac{\partial}{\partial x}\left(-k\frac{\partial T}{\partial x}\right)dxdy - \frac{\partial}{\partial x}(\varrho u c_p T)dxdy - \frac{\partial}{\partial y}\left(-k\frac{\partial T}{\partial y}\right)dydz$$
$$-\frac{\partial}{\partial y}(\varrho v c_p T)dydx = 0$$

$$k\frac{\partial^2 T}{\partial x^2} - \frac{\partial}{\partial x}(\varrho u c_p T) + k\frac{\partial^2 T}{\partial y^2} - \frac{\partial}{\partial y}(\varrho v c_p T) = 0,$$

$$k\left(\frac{\partial^2 T}{\partial x^2} + \frac{\partial^2 T}{\partial y^2}\right) - \varrho c_p \left[\frac{\partial}{\partial x}(uT) + \frac{\partial}{\partial y}(vT)\right] = 0,$$

$$k\left(\frac{\partial^2 T}{\partial x^2} + \frac{\partial^2 T}{\partial y^2}\right) - \varrho c_p \left[u\frac{\partial T}{\partial x}v + \frac{\partial T}{\partial y} + T\left(\frac{\partial u}{\partial x} + \frac{\partial v}{\partial y}\right)\right] = 0,$$

using the continuity equation, definition of Prandtl number and noting that

$$\frac{\partial^2 T}{\partial x^2} \ll \frac{\partial^2 T}{\partial y^2}$$

we get

$$u\frac{\partial T}{\partial x} + v\frac{\partial T}{\partial y} = \frac{\nu}{\Pr}\left(\frac{\partial^2 T}{\partial y^2}\right).$$

3.4 Problem

Since the governing equations for two-dimensional and axisymmetric flows differ from each other only by the radial distance $r(x,y)$, the axisymmetric flow equations, Eqs. (3.3.1) to (3.3.3) can be placed in a nearly two-dimensional form by using a transformation known as the *Mangler transformation*. In the case of flow over a body of revolution of radius r_0 (a function of x), we find that if the boundary-layer thickness is small compared with r_0, so that $r(x,y) = r_0(x)$, this transformation puts them exactly in two-dimensional form. We define the Mangler transformation by

$$d\bar{x} = \left(\frac{r_0}{L}\right)^{2K} dx, \quad d\bar{y} = \left(\frac{r}{L}\right)^{K} dy \tag{P3.1}$$

to transform an axisymmetric flow with coordinates (x,y), into a two-dimensional flow with coordinates $(\bar{x},\bar{y})$. In Eq. (P3.1) L is an arbitrary reference length. If a stream function in Mangler variables $(\bar{x},\bar{y})$ is related to a stream function ψ in (x,y) variables by

$$\bar{\psi}(\bar{x},\bar{y}) = \left(\frac{1}{L}\right)^{K} \psi(x,y),$$

then

(a) show that the relation between the Mangler transformed velocity components $\bar{u}$ and $\bar{v}$ in $(\bar{x},\bar{y})$ variables and the velocity components u and v in (x,y) variables is:

$$u = \bar{u}$$

$$v = \left(\frac{L}{r}\right)^{K}\left[\left(\frac{r_0}{L}\right)^{2K}\bar{v} - \frac{\partial \bar{y}}{\partial x}\bar{u}\right]. \tag{P3.2}$$

(b) By substituting from Eqs. (P3.2) into Eqs. (3.3.1)–(3.3.3), show that for laminar flows the Mangler-transformed continuity momentum and energy equations are:

$$\frac{\partial \bar{u}}{\partial \bar{x}} + \frac{\partial \bar{v}}{\partial \bar{y}} = 0$$

$$\bar{u}\frac{\partial \bar{u}}{\partial \bar{x}} + \bar{v}\frac{\partial \bar{u}}{\partial \bar{y}} = -\frac{1}{\varrho}\frac{dp}{d\bar{x}} + \nu\frac{\partial}{\partial \bar{y}}\left[(1+t)^{2K}\frac{\partial \bar{u}}{\partial \bar{y}}\right] \tag{P3.3}$$

$$\bar{u}\frac{\partial T}{\partial \bar{x}} + \bar{v}\frac{\partial T}{\partial \bar{y}} = \frac{k}{\varrho c_p}\frac{\partial}{\partial \bar{y}}\left[(1+t)^{2K}\frac{\partial T}{\partial \bar{y}}\right] \tag{P3.4}$$

where

$$t = -1 + \left(1 + \frac{2L\cos\phi}{r_0^2}\bar{y}\right)^{1/2}. \tag{P3.5}$$

Note that for $t = 0$, Eqs. (P3.3) and (P3.4) in the $(\bar{x},\bar{y})$ plane are exactly in the same form as those for two-dimensional flows in the (x,y) plane.

SOLUTION

a. With the Mangler transformation given by Eqs. (P3.1)

$$\bar{\Psi}(\bar{x},\bar{y}) = (1/L)^{K}\Psi(x,y)$$

and with the chain rule

3. Boundary-Layer Equations

$$u = \frac{1}{r^K}\frac{\partial \Psi}{\partial y} = \left(\frac{L}{r}\right)^K \frac{\partial \overline{\Psi}}{\partial \overline{y}}\frac{\partial \overline{y}}{\partial y} = \left(\frac{L}{r}\right)^K \left(\frac{r}{L}\right)^K \frac{\partial \overline{\Psi}}{\partial \overline{y}} = \overline{u}$$

$$v = -\frac{1}{r^K}\frac{\partial \Psi}{\partial x} = -\left(\frac{L}{r}\right)^K \left(\frac{\partial \overline{\Psi}}{\partial \overline{x}}\frac{\partial \overline{x}}{\partial x} + \frac{\partial \overline{\Psi}}{\partial \overline{y}}\frac{\partial \overline{y}}{\partial x}\right)$$

$$= \left(\frac{L}{k}\right)^K \left[\overline{v}\left(\frac{r_0}{L}\right)^{2K} - \overline{u}\frac{\partial \overline{y}}{\partial x}\right].$$

Here

$$\overline{u} = \frac{\partial \overline{\Psi}}{\partial \overline{y}}, \quad \overline{v} = -\frac{\partial \overline{\Psi}}{\partial \overline{x}}$$

b. With the chain rule,

$$\frac{\partial r^K u}{\partial x} = u\frac{\partial r^K}{\partial x} + r^K\left(\frac{\partial u}{\partial \overline{x}}\frac{\partial \overline{x}}{\partial x} + \frac{\partial u}{\partial \overline{y}}\frac{\partial \overline{y}}{\partial x}\right)$$

$$= \overline{u}\frac{\partial r^K}{\partial x} + r^K\left[\left(\frac{r_0}{L}\right)^{2K}\frac{\partial \overline{u}}{\partial \overline{x}} + \frac{\partial \overline{u}}{\partial \overline{y}}\frac{\partial \overline{y}}{\partial \overline{x}}\right]$$

$$\frac{\partial r^K v}{\partial y} = \frac{\partial}{\partial y}\left[L^K \overline{v}\left(\frac{r_0}{L}\right)^{2K} - L^K \overline{u}\frac{\partial \overline{y}}{\partial x}\right]$$

$$= L^K \left[\left(\frac{r_0}{L}\right)^{2K}\frac{\partial \overline{v}}{\partial \overline{y}}\frac{\partial \overline{y}}{\partial y} - \frac{\partial \overline{u}}{\partial \overline{y}}\frac{\partial \overline{y}}{\partial y}\frac{\partial \overline{y}}{\partial x} - \overline{u}\frac{\partial}{\partial x}\left(\frac{\partial \overline{y}}{\partial y}\right)\right]$$

$$= r^K\left(\frac{r_0}{L}\right)^{2K}\frac{\partial \overline{v}}{\partial \overline{y}} - r^K\frac{\partial \overline{u}}{\partial \overline{y}}\frac{\partial \overline{y}}{\partial x} - \overline{u}\frac{\partial r^K}{\partial x}.$$

The continuity equation expressed in the (x,y) coordinate system can be written as

$$\frac{\partial r^K u}{\partial x} + \frac{\partial r^K v}{\partial y} = 0 = r^K\left(\frac{r_0}{L}\right)^{2K}\left(\frac{\partial \overline{u}}{\partial \overline{x}} + \frac{\partial \overline{v}}{\partial \overline{y}}\right) = \frac{\partial \overline{u}}{\partial \overline{x}} + \frac{\partial \overline{v}}{\partial \overline{y}} = 0.$$

Similarly,

$$u\frac{\partial u}{\partial x} = \overline{u}\left(\frac{\partial \overline{u}}{\partial \overline{x}}\frac{\partial \overline{x}}{\partial x} + \frac{\partial \overline{u}}{\partial \overline{y}}\frac{\partial \overline{y}}{\partial x}\right) = \overline{u}\left[\left(\frac{r_0}{L}\right)^{2K}\frac{\partial \overline{u}}{\partial \overline{x}} + \frac{\partial \overline{u}}{\partial \overline{y}}\frac{\partial \overline{y}}{\partial x}\right]$$

$$v\frac{\partial u}{\partial y} = \left(\frac{L}{r}\right)^K\left[\overline{v}\left(\frac{r_0}{L}\right)^{2K} - \overline{u}\frac{\partial \overline{y}}{\partial x}\right]\frac{\partial \overline{u}}{\partial \overline{y}}\left(\frac{r}{L}\right)^K$$

$$\frac{dp}{dx} = \frac{dp}{d\overline{x}}\frac{d\overline{x}}{dx} = \frac{dp}{d\overline{x}}\left(\frac{r_0}{L}\right)^{2K}$$

$$\nu\frac{1}{r}\frac{\partial}{\partial y}\left(r\frac{\partial u}{\partial y}\right) = \nu\frac{1}{r}\frac{\partial}{\partial \overline{y}}\left[r\frac{\partial \overline{u}}{\partial \overline{y}}\left(\frac{r}{L}\right)\right]\frac{r}{L} = \nu\left(\frac{r_0}{L}\right)^2\frac{\partial}{\partial \overline{y}}\left[\left(\frac{r}{r_0}\right)^2\frac{\partial \overline{u}}{\partial \overline{y}}\right].$$

Noting the definition of $r = r_0 + y\cos\alpha$, and denoting $t = y\cos\alpha/r_0$

$$\left(\frac{r}{r_0}\right)^2 = \left(1 + \frac{y\cos\alpha}{r_0}\right)^2 = (1+t)^2.$$

To express t as a function of $\bar{y}$, we write the relation between $\bar{y}$ and y from $d\bar{y} = r/L\,dy = [(r_0 + y\cos\alpha)/L]dy$, $\bar{y} = (r_0/L)y + 1/2L\cos\alpha\, y^2$ and solve for y with the definition of t given by

$$t = \frac{y\cos\alpha}{r_0} = -1 + \sqrt{1 + \frac{2L\cos\alpha}{r_0^2}\bar{y}}.$$

Substituting the above expressions for $u(\partial u/\partial x)$ etc. into the momentum equation yields

$$\bar{u}\frac{\partial \bar{u}}{\partial x} + \bar{v}\frac{\partial \bar{u}}{\partial \bar{y}} = -\frac{1}{\varrho}\frac{\partial p}{\partial x} + \nu\frac{\partial}{\partial \bar{y}}\left[(1+t)^{2K}\frac{\partial \bar{u}}{\partial \bar{y}}\right].$$

Similarly, the energy equation can be written as

$$\bar{u}\frac{\partial T}{\partial x} + \bar{v}\frac{\partial T}{\partial \bar{y}} = \frac{k}{\varrho c_p}\frac{\partial}{\partial \bar{y}}\left[(1+t)^{2K}\frac{\partial T}{\partial \bar{y}}\right]$$

3.5 Problem

Show that if in the decelerating turbulent boundary layer in an expanding passage (a diffuser) the skin-friction term in the momentum-integral equation, Eq. (3.5.14), is negligible and H can be taken as constant, Eq. (3.5.14) gives

$$\frac{\theta}{\theta_0} = \left(\frac{u_e}{u_{e,0}}\right)^{-(H+2)}$$

where subscript 0 denotes initial conditions.

SOLUTION

If c_f is negligible, Eq. (3.5.24b) becomes

$$\frac{d\theta}{dx} = -\frac{\theta}{u_e}\frac{du_e}{dx}(H+2)$$

and integration with the assumption that H is constant, yields

$$\ln\theta = -(H+2)\ln u_e + \text{const},$$

so that

$$\frac{\theta}{\theta_0} = \left(\frac{u_e}{u_{e,0}}\right)^{-(H+2)}$$

where subscript 0 denotes initial conditions.

3. Boundary-Layer Equations

3.6 PROBLEM

Show that for an incompressible zero-pressure gradient flow over a wall at uniform temperature, Eq. (3.5.24b) can be written as

$$\frac{d\theta_T}{dx} = \text{St}. \tag{P3.6}$$

SOLUTION

For an incompressible zero pressure-gradient flow,

$$\frac{du_e}{dx} = 0$$

and Eq. (3.5.14) reduces to

$$\frac{d\theta_T}{dx} = \text{St}.$$

3.7 PROBLEM

While Eq. (3.2.10b) expresses conservation of momentum at each point within the boundary layer, Eq. (3.5.13) expresses conservation of momentum for the boundary layer as a whole. Another useful equation is the so-called kinetic-energy integral equation which expresses the physical fact that the rate of change of the flux of kinetic-energy defect within the boundary layer is equal to the rate at which kinetic energy is dissipated by viscosity. It is given by

$$\frac{d}{dx}\left(\frac{1}{2}\varrho u_e^3 \delta^{**}\right) = \mu \int_0^\infty \left(\frac{\partial u}{\partial y}\right)^2 dy. \tag{P3.7}$$

Here δ^{**} denotes the kinetic-energy thickness which measures the flux of kinetic energy defect within the boundary layer as compared with an inviscid flow. It is defined for constant-density flow by

$$\delta^{**} = \int_0^\infty \frac{u}{u_e}\left(1 - \frac{u^2}{u_e^2}\right) dy. \tag{P3.8}$$

Suggestion: Using the procedure outlined below, derive Eq. (P3.7).
(a) First multiply Eq. (3.2.10b) by u and integrate it across the layer.
(b) Integrate of the term involving v by "parts."
(c) Note that

$$\int_0^\infty u\frac{\partial^2 u}{\partial y^2} dy = -\int_0^\infty \left(\frac{\partial u}{\partial y}\right)^2 dy.$$

(d) Using the steps (b) and (c) in step (a), after multiplication by minus 2, show that the resulting expression can be put in the same form as Eq. (P3.7).

SOLUTION

Multiply Eq. (3.2.10b) by u and integrate it across the layer,

$$\int_0^\infty \varrho u^2 \frac{\partial u}{\partial x} dy + \int_0^\infty \varrho uv \frac{\partial u}{\partial y} dy = \int_0^\infty \varrho u u_e \frac{du_e}{dx} dy + \int_0^\infty \mu u \frac{\partial^2 u}{\partial y^2} dy \quad \text{(P3.1)}$$

and then evaluate each term

$$\int_0^\infty \varrho u^2 \frac{\partial u}{\partial x} dy = \frac{1}{2} \int_0^\infty \varrho u \frac{\partial u^2}{\partial x} dy = \frac{1}{2} \int_0^\infty \varrho \left(\frac{\partial u^3}{\partial x} - u^2 \frac{\partial u}{\partial x} \right) dy \quad \text{(2a)}$$

$$\int_0^\infty \varrho uv \frac{\partial u}{\partial y} dy = \frac{1}{2} \int_0^\infty \varrho v \frac{\partial u^2}{\partial y} dy = \frac{1}{2} \left(\varrho v u^2 \Big|_0^\infty - \int_0^\infty u^2 \frac{\partial v}{\partial y} dy \right)$$

$$= \frac{1}{2} \left(-u_e^2 \int_0^\infty \frac{\partial u}{\partial x} dy - \int_0^\infty u^2 \frac{\partial v}{\partial y} dy \right). \quad \text{(2a)}$$

With $v_e = -\int_0^\infty \partial u/\partial x \, dy$ from the continuity equation,

$$\int_0^\infty \varrho u u_e \frac{du_e}{dx} dy = \frac{1}{2} \int_0^\infty \varrho u \frac{du_e^2}{dx} dy \quad \text{(2c)}$$

$$\int_0^\infty \mu u \frac{\partial^2 u}{\partial y^2} dy = \mu u \frac{\partial u}{\partial y} \Big|_0^\infty - \mu \int_0^\infty \left(\frac{\partial u}{\partial y} \right)^2 dy. \quad \text{(2d)}$$

Substituting (2a)–(2d) into (1), we have

$$\frac{1}{2} \int_0^\infty \left(\frac{\partial u^3}{\partial x} - u^2 \frac{\partial u}{\partial x} \right) dy - \frac{1}{2} \int_0^\infty u_e^2 \frac{\partial u}{\partial x} dy - \frac{1}{2} \int_0^\infty u^2 \frac{\partial v}{\partial y} dy$$

$$- \frac{1}{2} \int_0^\infty \varrho u \frac{u_e^2}{dx} dy = -\mu \int_0^\infty \left(\frac{\partial u}{\partial y} \right)^2 dy.$$

After rearranging and using the continuity equation,

$$\frac{\partial u}{\partial x} + \frac{\partial v}{\partial y} = 0,$$

we obtain

$$\frac{1}{2} \int_0^\infty \varrho \frac{\partial}{\partial x} (u^3 - u u_e^2) dy = -\mu \int_0^\infty \left(\frac{\partial u}{\partial y} \right)^2 dy$$

which can be written in the same form as (P.3.7) by using the definition of the kinetic-energy thickness δ^* given by (P3.8).

4
Laminar Boundary Layers

4.1 Problem

Using the Falkner-Skan transformation given by Eqs. (4.3.1), show that the momentum and energy equations, Eqs. (3.2.10b) and (3.2.11) and their boundary conditions, Eqs. (3.6.1b) and (3.6.2b), can be expressed in the forms given by Eqs. (4.4.1) to (4.4.3). Assume that wall temperature is specified and use the definition of dimensionless temperature given by Eq. (4.3.6).

Solution

Using the Falkner-Skan transformation given Eqs. (4.3.1) and the definition of dimensionless temperature g given by (4.3.6) it follows from the chain rule that

$$u = \left(\frac{\partial \psi}{\partial y}\right)_x = \frac{\partial \psi}{\partial \eta}\frac{\partial \eta}{\partial y} = u_e f'$$

$$v = -\left(\frac{\partial \psi}{\partial x}\right)_y = -\left[\left(\frac{\partial \psi}{\partial x}\right)_\eta + \frac{\partial \psi}{\partial \eta}\frac{\partial \eta}{\partial x}\right]$$
$$= -(u_e \nu x)^{1/2}\left[\frac{\partial f}{\partial x} + \frac{1}{2}\left(\frac{1}{u_e}\frac{du_e}{dx} + \frac{1}{x}\right)f + f'\frac{\partial \eta}{\partial x}\right]$$

$$u\frac{\partial u}{\partial x} = u_e f'\left[\left(\frac{\partial u}{\partial x}\right)_\eta + \frac{\partial u}{\partial \eta}\frac{\partial \eta}{\partial x}\right] = u_e f'\left(\frac{du_e}{dx}f' + u_e\frac{\partial f'}{\partial x} + u_e f''\frac{\partial \eta}{\partial x}\right)$$

$$v\frac{\partial u}{\partial y} = v\frac{\partial u}{\partial \eta}\frac{\partial \eta}{\partial y} = -u_e^2\left[f''\frac{\partial f}{\partial x} + \frac{1}{2x}\left(\frac{x}{u_e}\frac{du_e}{dx} + 1\right)ff'' + f'f''\frac{\partial \eta}{\partial x}\right]$$

$$\nu\frac{\partial^2 u}{\partial y^2} = \frac{u_e^2}{x}f'''$$

$$u\left(\frac{\partial T}{\partial x}\right)_y = u_e f'\left[(1-g)\frac{d(T_w - T_e)}{dx} - (T_w - T_e)\frac{\partial q}{\partial x}\right]$$

$$-u_e(T_w - T_e)f'g'\frac{\partial \eta}{\partial x}$$

$$v\frac{\partial T}{\partial y} = u_e(T_w - T_e)\left[g'\frac{\partial f}{\partial x} + \frac{1}{2x}\left(\frac{x}{u_e}\frac{du_e}{dx} + 1\right)fg' + f'g'\frac{\partial \eta}{\partial x}\right]$$

$$\frac{\nu}{\Pr}\frac{\partial^2 T}{\partial y^2} = -\frac{T_w - T_e}{\Pr}\left(\frac{u_e}{\nu x}\right)g''.$$

Substituting the above relations into (3.2.10b) and (3.2.11) and, after some arrangement, we obtain the expressions given by (4.3.3) and (4.3.10) with m and n given by (4.3.5) and (4.3.11a). For similar flows, f and g are functions of η only and, as a result, (4.3.3) and (4.3.10) reduce to (4.4.1) and (4.4.2) subject to the conditions given by

$$\eta = 0: \quad f = f' = 0, \quad g = 0; \quad \eta \to \infty: \quad f' = 1, \quad g = 1.$$

4.2 Problem

The Von Mises transformation is sometimes used to place the boundary-layer equations into a more amenable form for solution. In this transformation new independent variables (x, ψ) are taken with ψ denoting the stream function

$$\psi = \int_0^y \varrho u \, dy$$

Show that with this transformation the momentum and energy equations given by Eqs. (3.2.10b) and (3.2.11), subject to the boundary conditions given by Eqs. (3.6.1b) and (3.6.2b) can be written as

$$u\frac{\partial u}{\partial x} - u_e\frac{du_e}{dx} = \nu u \frac{\partial}{\partial \psi}\left(u\frac{\partial u}{\partial \psi}\right) \tag{P4.1}$$

$$\frac{\partial T}{\partial x} = \frac{\nu}{\Pr}\frac{\partial}{\partial \psi}\left(u\frac{\partial T}{\partial \psi}\right) \tag{P4.2}$$

$$\psi = 0; \quad u = 0, \quad T = T_w \tag{P4.3a}$$

$$\psi \to \infty; \quad u \to u_e, \quad T \to T_e. \tag{P4.3b}$$

Note that when u is determined from Eqs. (P4.1) and (P4.3), then v follows from the continuity equation (3.2.1). Note also that with this transformation, momentum and energy equations are placed in the form of the nonlinear unsteady heat conduction equation,

$$\frac{\partial T}{\partial t} = \frac{\partial}{\partial x}\left[a(t, x)\frac{\partial T}{\partial x}\right].$$

4. Laminar Boundary Layers

The most useful applications of this transformation have been in laminar heat transfer problems.

SOLUTION

In the new coordinate system (x, ψ) the terms of (3.2.10b) and (3.2.11) with
$$u = \frac{\partial \psi}{\partial y}, \quad v = -\frac{\partial \psi}{\partial x},$$
can be expressed as

$$u\left(\frac{\partial u}{\partial x}\right)_y = u\left[\left(\frac{\partial u}{\partial x}\right)_\psi + \frac{\partial u}{\partial \psi}\frac{\partial \psi}{\partial x}\right] = u\frac{\partial u}{\partial x} - uv\frac{\partial u}{\partial \psi}$$

$$v\frac{\partial u}{\partial y} = v\frac{\partial u}{\partial \psi}\frac{\partial \psi}{\partial y} = uv\frac{\partial u}{\partial \psi}$$

$$\nu\frac{\partial^2 u}{\partial y^2} = \nu\frac{\partial}{\partial \psi}\left(\frac{\partial u}{\partial \psi}\frac{\partial \psi}{\partial y}\right)\frac{\partial \psi}{\partial y} = \nu u\frac{\partial}{\partial \psi}\left(u\frac{\partial u}{\partial \psi}\right)$$

$$u\frac{\partial T}{\partial x} = u\left[\left(\frac{\partial T}{\partial x}\right)_\psi + \frac{\partial T}{\partial \psi}\frac{\partial \psi}{\partial x}\right] = u\frac{\partial T}{\partial x} - uv\frac{\partial T}{\partial \psi}$$

$$v\frac{\partial T}{\partial y} = uv\frac{\partial T}{\partial \psi},$$

$$\frac{\nu}{\Pr}\frac{\partial^2 T}{\partial y^2} = \frac{\nu}{\Pr}u\frac{\partial}{\partial \psi}\left(u\frac{\partial T}{\partial \psi}\right).$$

Substituting the above relations into (3.2.10b) and (3.2.11) we obtain (P4.1) and (P4.2).

4.3 PROBLEM

(a) Show that if we integrate Eq. (4.4.1) across the boundary layer, use the definitions of dimensionless displacement and momentum thickness, δ_1^* and θ_1 (see Eqs (4.4.7) and (4.4.8) respectively), and note that f'' is zero at the edge of the boundary layer because $\partial u/\partial y$ is zero there, we get, for no mass transfer,

$$-f''_w + \frac{m+1}{2}\int_0^{\eta_e} ff''\, d\eta + m(\delta_1^* + \theta_1) = 0. \tag{P4.4}$$

(b) Show that Eq (P4.4) can be written as

$$-f''_w + m\delta_1^* + \left(\frac{3m+1}{2}\right)\theta_1 = 0 \tag{P4.5}$$

by noting that the integral in Eq. (P4.4) is the dimensionless momentum thickness θ_1 defined by Eq. (4.4.8). Note that Eq. (P4.5) provides a relation between the three boundary-layer parameters, f''_w, θ_1, and δ_1^* for a given m. In this way θ_1 can be obtained exactly without evaluating Eq. (4.4.8) numerically.

SOLUTION

a. Integrating (4.4.1) with respect to η from 0 to η_e, we get

$$\int_0^{\eta_e} f''' d\eta + \frac{m+1}{2} \int_0^{\eta_e} f f'' d\eta + m \int_0^{\eta_e} (1 - f'^2) d\eta = 0. \quad (1)$$

Noting that

$$\int_0^{\eta_e} f''' d\eta = f'' \Big|_0^{\eta_e} = -f''_w$$

$$m \int_0^{\eta_e} (1 - f'^2) d\eta = m \int_0^{\eta_e} [1 - f' + (1 - f')] d\eta = m(\delta_1^* + \theta_1).$$

Eq. (1) can be written in the form given by (P4.4).

b. Integrating by parts,

$$\int_0^{\eta_e} f f'' d\eta = f f' \Big|_0^{\eta_e} - \int_0^{\eta_e} [f' - f'(1 - f')] d\eta$$

$$= f_e - (f_e - f_w - \theta_1) = f_w + \theta_1$$

where f_w is the value of f at the wall and is zero when there is no mass transfer. Thus, $\int_0^{\eta_e} f f'' d\eta = \theta_1$ and (P4.4) reduces to that given by (P4.5).

4.4 PROBLEM

As in similar velocity layers, we can also obtain useful integral relations for similar thermal layers for uniform wall temperature. Setting $n = 0$ and assuming no mass transfer at the surface, show that Eq. (4.4.2) can be integrated with respect to η from $\eta = 0$ to $\eta = \eta_e$ for $g'(\eta_e) = 0$, $g(\eta_e) = 1$, to get

$$-g'_w + \Pr \frac{m+1}{2} \left[f(\eta_e) - \int_0^{\eta_e} f' g \, d\eta \right] = 0. \quad (P4.6)$$

Show that Eq. (P4.6) can be written as

$$-g'_w + \Pr \frac{m+1}{2} \theta_{T_1} = 0 \quad (P4.7)$$

where

$$\theta_{T_1} = \int_0^{\eta_e} f'(1 - g) \, d\eta.$$

For similar thermal boundary layers with uniform wall temperature, this equation can be used to compute θ_T without evaluating Eq. (3.5.26) numerically.

4. Laminar Boundary Layers

SOLUTION

For constant wall temperature, (4.4.2) reduces to (4.4.14). Integrating the first term of this equation with respect to η from 0 η_e and using the relations
$$\int_0^{\eta_e} g'' d\eta = (g'_e - g'_w) = -g'_w$$
and
$$\Pr\frac{m+1}{2}\int_0^{\eta_e} fg' d\eta = \Pr\frac{m+1}{2}\left[(fg)_0^{\eta_e} - \int_0^{\eta_e} gf' d\eta\right]$$
$$= \Pr\frac{m+1}{2}(f_e - f_e + f_w + \theta_T)$$
we get (P4.7) with θ_{T_1} as defined in Problem (P4.6).

4.5 PROBLEM

Show that for a similar velocity and thermal boundary-layer flow with porous wall and nonuniform wall temperature, Eqs. (4.4.1) and (4.4.2) can be written as

$$-f''_w + \frac{m+1}{2}f_w + m\delta_1^* + \frac{3m+1}{2}\theta_1 = 0 \qquad (P4.8)$$

$$-g'_w + \frac{1}{2}\Pr(m+1)f_w + \Pr\left(\frac{m+1}{1} + n\right)\theta_{T_1} = 0. \qquad (P4.9)$$

SOLUTION

Integrating (4.4.1) and (4.4.2) wrt η from 0 to η_e, we get

$$-f''_w + \frac{m+1}{2}\int_0^{\eta_e} ff'' d\eta + m(\delta_1^* + \theta_1) = 0, \qquad (1)$$

$$\frac{-g'_w}{\Pr} + \frac{m+1}{2}\int_0^{\eta_e} fg' d\eta + n\int_0^{\eta_e} f'_1(1-g) d\eta = 0. \qquad (2)$$

Integration by parts yields
$$\frac{m+1}{2}\int_0^{\eta_e} ff'' d\eta = \frac{m+1}{2}\left[ff'\Big|_0^{\eta_e} - \int_0^{\eta_e} [f' - f'(1-f')] d\eta\right]$$
$$= \frac{m+1}{2}(f_w + \theta_1) \qquad (3)$$

$$\frac{m+1}{2}\int_0^{\eta_e} fg' d\eta = \frac{m+1}{2}\left[fg\Big|_0^{\eta_e} - \int_0^{\eta_e} [f' - f'(1-g)] d\eta\right]$$
$$= \frac{m+1}{2}\left[f_e - f_e + f_w + \int_0^{\eta_e} f'(1-g) d\eta\right] = \frac{m+1}{2}(f_w + \theta_T). \qquad (4)$$

Using the definitions of θ_1 amd θ_T and substituting (3) into (1) and (4) into (2), we get the relations given by (P4.8) and (P4.9).

4.6 PROBLEM

Glycerine at $40\,°\mathrm{C}$ flows at a velocity of $3\,\mathrm{m\,s^{-1}}$ past a $4\,\mathrm{m}$ long flat plate whose surface is maintained at $20\,°\mathrm{C}$. Calculate per unit width, the heat transferred to the plate.

SOLUTION

With $\varrho = 1258\,\mathrm{kg/m^3}$, $\nu = 0.5 \times 10^{-3}\,\mathrm{m^2/s}$, $c_p = 2443.0\,\mathrm{J/kg\,K}$ and $\mathrm{Pr} = 5380$ for glycerine at $T_f = \frac{1}{2}(T_w + T_e) = 30\,°\mathrm{C}$, and since $\mathrm{Pr} \ll 1$, $\mathrm{Nu}_x = 0.339\mathrm{Pr}^{1/3}R_x^{1/2}$ and the local heat transfer rate is then (from the plate)

$$\dot{q}_w = \mathrm{St}\varrho c_p (T_w - T_e) u_e = \frac{\mathrm{Nu}_x}{\mathrm{Pr}R_x}\varrho c_p (T_w - T_e)u_e$$

$$= 0.339\mathrm{Pr}^{-2/3}R_x^{-1/2}\varrho c_p (T_w - T_e)u_e.$$

The total heat transfer to the plate is

$$q_w = -\int_0^L \dot{q}_w dx = 0.339\mathrm{Pr}^{-2/3}\varrho c_p(T_e - T_w)u_e\frac{2L}{\sqrt{R_L}} = 0.339(5380)^{2/3}$$

$$\times 1258 \times 2443 \times (40-20) \times 3 \times \frac{2\times 4}{\sqrt{(3\times 4)/(0.5\times 10^{-3})}}$$

$$= 1.0513 \times 10^4\,\mathrm{W/m}.$$

4.7 PROBLEM

Air at standard temperature and pressure flows normal to a uniformly heated $2\,\mathrm{cm}$ diameter circular cylinder at a velocity of $1\,\mathrm{m\,s^{-1}}$.

(a) Show that the Nusselt number based on cylinder radius is

$$\mathrm{Nu} = \sqrt{2}g'_w\sqrt{R}$$

where

$$R = \frac{u_\infty r_0}{\nu}.$$

(b) Taking $\mathrm{Pr} = 0.72$, $\nu = 1.5 \times 10^{-5}\,\mathrm{m^2\,s^{-1}}$, calculate the displacement thickness δ^* and the enthalpy thickness θ_T at the stagnation point.

SOLUTION

a. The Nusselt number based on the cylinder radius is

$$\mathrm{Nu}_{r_0} = \frac{r_0 \dot{q}_w}{k(T_w - T_e)} = -\frac{r_0(\partial T/\partial y)_w}{T_w - T_e}.$$

When the wall temperature is specified, we define the dimensionless temperature, g, by (4.3.6) and the similarity parameter η by (4.3.1a) and write

$$\left(\frac{\partial T}{\partial y}\right)_w = (T_e - T_w)\frac{\partial g}{\partial \eta}\frac{\partial \eta}{\partial y} = (T_e - T_w)\sqrt{\frac{u_e}{\nu x}}g'_w$$

so that the expressions for the Nusselt number, with $u_e = (2u_\infty x/r_0)$ and $R = (u_\infty r_0/\nu)$, becomes

$$\mathrm{Nu}_{r_0} = -\frac{r_0}{T_w - T_e}(T_e - T_w)\sqrt{\frac{u_e}{\nu x}}g'_w = r_0\sqrt{\frac{2u_\infty x}{r_0}\frac{1}{\nu x}}g'_w = \sqrt{2}\sqrt{R}g'_w.$$

b. With δ^* and θ_T representing the displacement and enthalpy thicknesses,

$$\delta^* = x\delta_1^*/\sqrt{R_x}, \quad \theta_T = x\theta_{T_1}/\sqrt{R_x}$$

where

$$\delta_1^* = \int_0^\infty (1-f')d\eta, \quad \theta_{T_1} = \int_0^\infty f'(1-g)d\eta.$$

At a stagnation point, $m = 1$ and $\delta_1^* = 0.64791$, $\theta_{T_1} = g'_w/\mathrm{Pr} = 0.5017/0.72 = 0.697$, so that with $u_e = 2u_\infty x/r_0$,

$$\delta^* = \frac{x}{\sqrt{2u_\infty x/r_0 \, x/\nu}}\delta_1^* = r_0/\sqrt{2u_\infty r_0/\nu}\,\delta_1^*$$

$$= 0.01/\sqrt{\frac{2\times 1.0 \times 0.01}{1.5\times 10^{-5}}} \times 0.64791 = 0.177\,\mathrm{mm}$$

$$\theta_T = \frac{r_0}{\sqrt{2u_\infty r_0/\nu}}\theta_{T_1} = 0.01/\sqrt{\frac{2\times 1.0\times 0.01}{1.5\times 10^{-5}}}\times 0.697 = 0.191\,\mathrm{mm}.$$

4.8 PROBLEM

Derive Eq. (4.4.26). (See Appendix C.)

SOLUTION

To derive (4.4.26), we start with (4.4.18) which is valid for all Pr and m. For $\mathrm{Pr} \gg 1$, take $f' = f''_w \eta$ so that $f = \eta^2/2f''_w$. From the definition of the Nusselt number

$$\mathrm{Nu} = \left(\frac{\partial q}{\partial \eta}\right)_w \sqrt{R_x} = R_x^{1/2}\left\{\int_0^{\eta_e}\exp\left[-\mathrm{Pr}(m+1)/2\int_0^\zeta f(z)dz\right]d\zeta^{-1}\right\}^{-1}$$

and from the expression for f, it follows that

$$\int_0^{\eta_e}\exp\left[-\mathrm{Pr}\frac{m+1}{2}\int_0^\zeta f(z)dz\right]d\zeta = \int_0^\infty \exp\left(-\mathrm{Pr}\frac{m+1}{12}f''_w\zeta^3\right)d\zeta. \tag{1}$$

Let $t = \mathrm{Pr}\frac{m+1}{12}f''_w\zeta^3$ so that $\frac{1}{3}(\mathrm{Pr}\frac{m+1}{2}f''_w)^{1/3}t^{-2/3}dt = d\zeta$. The RHS of (1) becomes

$$\int_0^\infty \exp\left(-\Pr\frac{m+1}{12}f_w''\zeta^3\right)d\zeta = \frac{1}{3}\Pr{}^{-1/3}\left(\frac{m+1}{12}f_w''\right)^{-1/3}\int_0^\infty e^{-t}t^{(1/3)-1}dt$$

$$= \frac{1}{3}\Gamma(1/3)\Pr{}^{-1/3}\left(\frac{m+1}{12}f_w''\right)^{-1/3}$$

so that with $\Gamma(1/3) = 2.6789$,

$$\mathrm{Nu} = R_x^{1/2}\Pr{}^{1/3}\left(\frac{m+1}{12}f_w''\right)^{1/3}\frac{3}{\Gamma(1/3)} = 1.12\left(\frac{m+1}{12}f_w''\right)^{1/3}R_x^{1/2}\Pr{}^{1/3}.$$

4.9 PROBLEM

When there is a jump in the wall temperature distribution so that $\partial T/\partial x$ is discontinuous, say at $x = x_0$, a solution of the energy equation in the neighborhood of x_0 can be obtained analytically by using expansions of u, v and T. Let us assume that the velocity field is known and that we are interested in the solution of the energy equation with g given by Eq. (4.3.6). The energy equation

$$u\frac{\partial g}{\partial x} + v\frac{\partial g}{\partial y} = \frac{\nu}{\Pr}\frac{\partial^2 g}{\partial y^2} \tag{P4.10}$$

is subject to

$$y = 0, \quad g = 0 \text{ for } x < x_0, \quad g = 1 \text{ for } x > x_0 \tag{P4.11a}$$

$$y = \delta_t, \quad g = 0 \quad x > x_0. \tag{P4.11b}$$

For $x > x_0$, we assume that u and v do not change much in the x-direction in the neighborhood of the discontinuity and we expand them by

$$u = \sum_{k=1}^\infty \lambda_k y^k, \quad v = \sum_{k=2}^\infty \mu_k y^k. \tag{P4.12}$$

We expand the dimensionless g by

$$g = \sum_{n=0}^\infty z^{n/3}G_n(\zeta) \tag{P4.13}$$

where

$$z = x - x_0, \quad \zeta = y/z^{1/3}. \tag{P4.14}$$

(a) Show that for $n = 0$ with primes denoting differentiation with respect to ζ and $\lambda_1 = (\partial u/\partial y)_w$, Eq. (P4.10) can be written as

$$G_0'' + \frac{1}{3}\lambda_1\frac{\Pr}{\nu}\zeta^2 G_0' = 0. \tag{P4.15}$$

4. Laminar Boundary Layers

(b) Show tat the solution of Eq. (P4.15) is

$$G_0(\zeta) = 1 + A \int_0^\zeta \left(-\frac{1}{9}\lambda_1 \frac{\Pr}{\nu} \zeta^3\right) d\zeta \quad \text{(P4.16)}$$

where

$$A = -\frac{3}{\Gamma(1/3)} \left(\frac{\Pr \lambda_1}{9\nu}\right)^{1/3} \quad \text{(P4.17)}$$

and the Gamma function (see Appendix C) $\Gamma(1/3)$ is equal to 2.679 approximately.

(c) Show that for a flat-plate laminar flow

$$\lambda_1 = 0.332 \frac{u_e}{x_0} \sqrt{R_{x_0}} \quad \text{(P4.18)}$$

where

$$R_{x_0} = \frac{u_e x_0}{\nu}.$$

(d) Show that

$$\mathrm{Nu}_{x_0} = C R_{x_0}^{1/2} \Pr^{1/3} \frac{1}{[(x/x_0) - 1]^{1/3}} \quad \text{(P4.19)}$$

where

$$\mathrm{Nu}_{x_0} = \frac{\dot{q}_w}{(T_w - T_e)} \frac{x_0}{k}$$

$$C = \frac{3}{\Gamma(1/3)} \left(\frac{0.332}{9}\right)^{1/3}.$$

SOLUTION

a. Consider the energy equation, (P4.10) and its boundary conditions $y = 0 : g = 0$ for $x < x_0$, $g = 1$ for $x > x_0$ and $y = \delta_t : g = 0$ for $x > x_0$. Define a new variable ζ such that

$$\zeta = y/z^{1/3}, \quad z = x - x_0$$

so that

$$u = \sum_{k=1}^\infty \lambda_k y^k = \sum_{k=1}^\infty \lambda_k z^{k/3} \zeta^k,$$

$$v = \sum_{k=2}^\infty \mu_k y^k = \sum_{k=2}^\infty \mu_k z^{k/3} \zeta^k,$$

$$g = \sum_{k=0}^\infty z^{k/3} G_k(\zeta)$$

chain rule gives

$$\left(\frac{\partial q}{\partial x}\right)_y = \left(\frac{\partial q}{\partial x}\right)_\zeta - \frac{1}{3}\frac{\zeta}{z}g', \quad \left(\frac{\partial q}{\partial y}\right)_x = z^{-1/3}g', \quad \left(\frac{\partial^2 q}{\partial y^2}\right) = z^{-2/3}g''.$$

Substituting the relations into (P4.10) yields

$$\sum_{k=1}^\infty \lambda_k z^{k/3}\zeta^k \left[\sum_{k=0}^\infty z^{n/3}\left(z\frac{\partial G_n}{\partial x} - \frac{1}{3}\zeta G_n'\right)\right]$$
$$+ z^{2/3}\left(\sum_{k=2}^\infty \mu z^{k/3}\zeta^k\right)\left(\sum_{k=0}^\infty z^{n/3}G_n'\right) = \frac{\nu}{\Pr}z^{1/3}\sum_{k=0}^\infty z^{n/3}G_n''.$$

Now expand the above expression and consider the lowest order of z, i.e. for $k = 1$ and $n = 0$,

$$-\frac{1}{3}\lambda_1\zeta^2 G_0' = \frac{\nu}{\Pr}G_0'' \quad \text{or} \quad G_0'' + \frac{1}{3}\lambda_1\frac{\Pr}{\nu}\zeta^2 G_0' = 0. \tag{P4.15}$$

b. Integrate (P4.15) twice with respect to ξ.

$$G_0 = b + a\int_0^\zeta \exp\left(-\frac{1}{g}\lambda_1\frac{\Pr}{\nu}\zeta^3\right)d\zeta.$$

The integration constants a and b are determined from

$$\zeta = 0, \quad G_0 = 1, \quad \zeta \to \infty, \quad G_0 \to 0$$

from which

$$b = 1.0, \quad a = -\frac{3}{\Gamma(1/3)}\left(\frac{1}{g}\lambda_1\frac{\Pr}{\nu}\right)^{1/3}.$$

Hence

$$G_0 = 1 - \frac{3}{\Gamma(1/3)}\left(\frac{1}{g}\lambda_1\frac{\Pr}{\nu}\right)^{1/3}\int_0^\zeta \exp\left(-\frac{1}{g}\lambda_1\frac{\Pr}{\nu}\zeta^3\right)d\zeta. \tag{P4.16}$$

c. With

$$u = \sum_{k=1}^\infty \lambda_k y^k, \quad \frac{\partial u}{\partial y} = \lambda_1 + \sum_{k=2}^\infty k\lambda_k y^{k-1}.$$

At

$$y = 0, \quad \lambda_1 = \left(\frac{\partial u}{\partial y}\right)_w = f_w''\frac{u_e}{x}\sqrt{\frac{u_e x}{\nu}}.$$

For a flat-plate laminar flow at $x = x_0$ with $f_w'' = 0.332$

$$\lambda_1 = 0.332\frac{u_e}{x_0}\sqrt{\frac{u_e x_0}{\nu}}$$

4. Laminar Boundary Layers

d.
$$\dot{q}_w = -k\left(\frac{\partial T}{\partial y}\right)_w = -k(T_w - T_e)\left(\frac{\partial G_0}{\partial \zeta}\right)_w \frac{\partial \zeta}{\partial y}$$
$$= \frac{k}{z^{1/3}}(T_w - T_e)\frac{3}{\Gamma(1/3)}\left(\frac{1}{9}\lambda_1\frac{\Pr}{\nu}\right)^{1/3}$$

from part (a)
$$\mathrm{Nu}_{x_0} = \frac{\dot{q}_w}{T_w - T_e}\frac{x_0}{k} = \frac{x_0}{z^{1/3}}\frac{3}{\Gamma(1/3)}\left(\frac{1}{9}\lambda_1\frac{\Pr}{\nu}\right)^{1/3}$$
$$= \frac{3}{\Gamma(\nu)}\left(\frac{\Pr}{\nu} \times \frac{0.332}{9}\frac{u_e}{x_0}\sqrt{\frac{u_e x_0}{\nu}}\right)^{1/3}\frac{x_0}{(x - x_0)^{1/3}}$$
$$= cR_{x_0}^{1/2}\Pr^{1/3}/\left(\frac{x}{x_0} - 1\right)^{1/3}$$

where
$$c = \frac{3}{\Gamma(1/3)}\left(\frac{0.332}{9}\right)^{1/3}, \qquad R_{x_0} = \frac{u_e x_0}{\nu}.$$

4.10 PROBLEM

Derive Eq. (4.5.11).
Hint: Let $s^3 = y$.

SOLUTION

Write (4.5.10) as
$$S^3 + \frac{4}{3}x\frac{dS^3}{dx} = \frac{13}{14}\frac{1}{\Pr}.$$

With
$$Y = S^3 - \frac{13}{14}\frac{1}{\Pr}$$

we have
$$Y + \frac{4}{3}x\frac{dY}{dx} = 0$$

so that
$$Y = cx^{-3/4} \quad \text{or} \quad S^3 = \frac{13}{14}\Pr^{-1} + cx^{-3/4}.$$

At
$$x = x_0, \quad S = 0, \quad 0 = cx_0^{-3/4} + \frac{13}{14}\Pr^{-1}.$$
$$\therefore \quad c = -\frac{13}{14}\Pr^{-1}x_0^{3/4}.$$
$$\therefore \quad S^3 = \frac{13}{14}\Pr^{-1}[1 - (x_0/x)^{3/4}] \quad \text{or} \quad S = \frac{1}{1.026}\Pr^{-1}[1 - (x_0/x)^{3/4}]^{1/3}.$$

4.11 PROBLEM

Consider an incompressible laminar flow over a flat plate which is unheated for $0 \leq x \leq x_0$ but for $x > x_0$, the plate is subjected to a constant heat flux. Using Eq. (3.5.23) and with velocity profile approximated by Eq. (4.5.4) and with the temperature profile approximated by

$$\frac{T - T_e}{T_w - T_e} = 1 - \frac{3}{2}\frac{y}{\delta_t} + \frac{1}{2}\left(\frac{y}{\delta_t}\right)^3$$

show that

(a) $\quad \mathrm{Nu}_x = \dfrac{\dot{q}_w x}{(T_w - T_e)k} = 0.417 R_x^{1/2}\mathrm{Pr}^{1/3}\left(1 - \dfrac{x_0}{x}\right)^{-1/3}.$ (P4.20)

(b) $\quad T_w - T_e = 2.40\dot{q}_w \dfrac{x}{k} R_x^{-1/2}\mathrm{Pr}^{-1/3}\left(1 - \dfrac{x_0}{x}\right)^{1/3}.$ (P4.21)

Hint: Note that the integration of Eq. (3.5.23) yields

$$u_e(T_w - T_e)\Theta_T = \frac{\dot{q}_w}{\varrho c_p}(x - x_0).$$

SOLUTION

a. Integrate the enthalpy integral equation,

$$\frac{d}{dx}[u_e \Theta_T(T_w - T_e)] = \frac{\dot{q}_w}{\varrho c_p}$$

wrt x to get

$$u_e \Theta_T(T_w - T_e) = \frac{x\dot{q}_w}{\varrho c_p} + c \quad (1)$$

at $x = x_0$, $\Theta_T = 0$, so $c = -\dfrac{\dot{q}_w x_0}{\varrho c_p}$.

$$\Theta_T = \frac{\dot{q}_w(x - x_0)}{\varrho c_p u_e(T_w - T_e)}. \quad (2)$$

To find Θ_T, use the velocity and temperature and temperature profiles given by

$$\frac{u}{u_e} = \frac{3}{2}\frac{y}{\delta} - \frac{1}{2}\left(\frac{y}{\delta}\right)^3, \quad \frac{T - T_e}{T_w - T_e} = 1 - \frac{3}{2}\frac{y}{\delta_t} + \frac{1}{2}\left(\frac{y}{\delta_t}\right)^3$$

which, with $S = \delta_t/\delta$, results in

$$\Theta_T = \delta\left(\frac{3}{20}S^2 - \frac{3}{280}S^4\right). \quad (3)$$

Since the second term in (3) is small compared with the first term, Θ_T can be approximated as

4. Laminar Boundary Layers

$$\theta_T = \frac{3}{20}\delta_t^2/\delta,$$

with δ_t given by

$$\delta_t = \frac{3k(T_w - T_e)}{\dot{q}_w}. \qquad (4)$$

From (2) and (4)

$$\frac{\dot{q}_w^3}{k^3(T_w - T_e)^3} = \frac{27}{80 \times 4.64} R_x^{3/2} \mathrm{Pr}\left(1 - \frac{x_0}{x}\right)^{-1}/x^3.$$

$$\therefore \ \mathrm{Nu}_x \equiv \frac{\dot{q}_w x}{k(T_w - T_e)} = 0.4174 R_x^{1/2} \mathrm{Pr}^{1/3}\left(1 - \frac{x_0}{x}\right)^{-1/3}.$$

b. From part (a),

$$T_w - T_e = \frac{x}{k}\dot{q}_w \left[0.417 R_x^{1/2} \mathrm{Pr}^{1/3}\left(1 - \frac{x_0}{x}\right)^{-1/3}\right]^{-1}$$

$$= 2.40 \dot{q}_w \frac{x}{k} R_x^{-1/2} \mathrm{Pr}^{-1/3}\left(1 - \frac{x_0}{x}\right)^{1/3}$$

4.12 PROBLEM

When the surface temperature of a flat plate is an arbitrary function of x, the energy equation can be solved accurately and easily for specified values of dimensionless temperature g_w by using the Fortran program given in Section 11.8. Approximate solutions for constant-property flow over a flat plate with non-uniform wall temperature can also be obtained by taking into account the fact that the energy equation for constant-property flow is a linear one so that we may use the superposition law for linear differential equations. We represent the wall temperature variation as one consisting of a sum of step functions ΔT_{wi} shown below and compute the total heat flux $\dot{q}_w$, from

$$\dot{q}_w = \sum_i \hat{h}_i \Delta T_{wi}. \qquad (\mathrm{P}4.22)$$

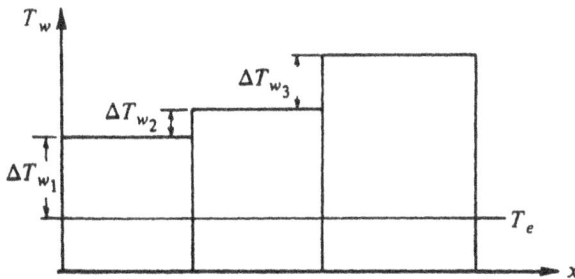

Problem 4.12. Stepwise variation of wall temperature.

Here $\hat{h}_i$ denotes the heat transfer coefficient due to a step function in wall temperature at x_i [see Eq. (4.5.15)] and can be written as

$$\hat{h}_i = \frac{0.332k\,\mathrm{Pr}^{1/3}\sqrt{R_x}}{x[1-(x_i/x)^{3/4}]^{1/3}} \qquad x > x_i. \tag{P4.23}$$

Using this property of the energy equation assume that the plate has the following temperature distribution: 0 to 0.5 m, 50 °C, 0.5 to 1 m, 60 °C, 1.0 to 1.5 m, 80 °C. Find the local heat flux at $x = 1.5$ m.
Note that in this problem $x_1 = 0$, $x_2 = 0.5$ m and $x_3 = 1.0$ and the first step in temperature is $T_{w1} - T_e$.

SOLUTION

$$\dot{q}_w = \sum_{i=1}^{3} \hat{h}_i \Delta T_{w_i} = \sum_{i=1}^{3} \frac{0.332k\mathrm{Pr}^{1/3}\sqrt{R_x}}{x[1-(x_i/x)^{3/4}]^{1/3}} \Delta T_{w_i}$$

$$= \frac{0.332k\mathrm{Pr}^{1/3}\sqrt{R_x}}{x} \times$$

$$\times \left[\frac{\Delta T_{w_1}}{[1-(x_1/x)^{3/4}]^{1/3}} + \frac{\Delta T_{w_2}}{[1-(x_2/x)^{3/4}]^{1/3}} + \frac{\Delta T_{w_3}}{[1-(x_3/x)^{3/4}]^{1/3}}\right].$$

With $x_1 = 0.0$, $x_2 = 0.5$, $x_3 = 1.0$, $x = 1.5$, $\Delta T_{w_1} = 50 - 20 = 30$, $\Delta T_{w_2} = 60 - 50 = 10$, $\Delta T_{w_3} = 80 - 60 = 20$, $k = 0.028$ W/m K, $\nu = 18.2 \times 10^{-6}$ m^2/s (for $T_m = 50\,°C$) $R_x = \frac{u_e x}{\nu} = 8.2 \times 10^5$.

$$\dot{q}_w = \frac{0.332 \times 0.028 \times (0.72)^{1/3}(8.3 \times 10^5)^{1/2}}{1.5}$$
$$\left(30.0 + \frac{10.0}{[1-(0.5/1.5)^{3/4}]^{1/3}} + \frac{20.0}{[1-(1.0/1.5)^{3/4}]^{1/3}}\right)$$
$$= 371\,\mathrm{W/m^2}.$$

4.13 PROBLEM

Show that Thwaites' formula, in its general form Eq. (4.5.19), follows if the velocity profile shape in a laminar boundary layer is uniquely related to the pressure gradient parameter $(\theta^2/\nu)du_e/dx$.

Hint: This implies that the velocity gradient at the wall, suitably non-dimensionalized, is a function of the pressure-gradient parameter.

SOLUTION

The integral momentum equation

$$\tau_w + \frac{dp}{dx}\delta^* = \frac{d}{dx}(\varrho u_e^2 \theta) \tag{1}$$

states that the total force acting on the fluid body is equal to the rate of change of the momentum flux. The last term in (1) represents the rate of the momentum flux. Therefore, the term

$$\frac{\delta^*}{\tau_w}\left(\frac{dp}{dx}\right) \tag{2}$$

represents the ratio of the pressure force to the wall shear force, and can be written as

$$\frac{\delta^*}{\tau_w}\frac{dp}{dx} = -\frac{\delta^*/u_e(du_e/dx)}{\tau_w/\varrho u_e^2}. \tag{3}$$

According to Thwaites' method,

$$\frac{c_f}{2} = \frac{\tau_w}{\varrho u_e^2} = \frac{\nu l(\lambda)}{u_e \theta}, \quad \lambda = \frac{\theta^2}{\nu}\frac{du_e}{dx}, \quad H = H(\lambda). \tag{4}$$

Substituting (4) into (3) yields

$$\frac{\delta^*}{\tau_w}\frac{dp}{dx} = -\frac{H\theta^2 du_e/dx}{\nu l(\lambda)} = -\lambda H(\lambda)/l(\lambda)$$

which shows that, to the accuracy of Thwaites' method, the ratio of two net forces is uniquely related to the pressure gradient parameter $\lambda = \theta^2/\nu du_e/dx$.

4.14 PROBLEM

Equation (4.5.31) is based on the assumption given by Eq. (4.5.25). For a Prandtl number equal to 10, determine the relationship

$$\frac{u_e}{\nu}\frac{d\delta_c^2}{dx} = F\left(\frac{\delta_c^2}{\nu}\frac{du_e}{dx}\right)$$

for wedge flows, $u_e = Cx^m$. That is, derive an expression for each of the terms in the above equation and indicate how you could make a plot which would describe the relationship between them.

Hint: For wedge flows, evaluate $u_e/\nu(d\delta_c/dx)$ and $\delta_c^2/\nu(du_e/dx)$ in terms of m and g'_w ($= \mathrm{Nu}_x/\sqrt{R_x}$). Note that $\delta_c/x = \mathrm{Nu}_x$.

SOLUTION

$$\delta_c^2 = \frac{x^2}{\mathrm{Nu}_x^2} = \frac{x^2}{R_x(g'_w)^2} = \frac{x\nu}{(g'_w)^2 u_e}$$

where g'_w is a function of Prandtl number and pressure gradient. For a fixed Pr and wedge flow, $u_e = cx^m$, g'_w is constant.

$$\therefore \frac{u_e}{\nu}\frac{d\delta_c^2}{dx} = \frac{u_e}{(g'_w)^2 \nu}\frac{d}{dx}\left(\frac{\nu}{c}x^{1-m}\right) = \frac{1-m}{(g'_w)^2},$$

$$\frac{\delta_c^2}{\nu}\frac{du_e}{dx} = \frac{x\nu}{(g'_w)^2 u_e \nu} cmx^{m-1} = \frac{m}{(g'_w)^2}$$

and as a result

$$\frac{u_e}{\nu}\frac{d\delta_c^2}{dx} = \frac{1-m}{m}\frac{\delta_c^2}{\nu}\frac{du_e}{dx}$$

for wedge flows. From this relationship, the two parameters $u_e/\nu\, d\delta_c^2/dx$ and $\delta_c^2/\nu\, du_e/dx$ can be linearly related

$$\frac{u_e}{\nu}\frac{d\delta_c^2}{dx} = A\left(\frac{\delta_c^2}{\nu}\frac{du_e}{dx}\right) + B$$

with the constants A and B depending on the Prandtl number.

4.15 Problem

Heat transfer from an isothermal flat plate can be obtained by an approximate procedure due to Leveque. This procedure considers the solution of the energy equation (3.2.11) in the region near the wall with Eq. (3.2.11) approximated to

$$u\frac{\partial T}{\partial x} = \frac{\nu}{\Pr}\frac{\partial^2 T}{\partial y^2}. \tag{P4.24}$$

It also assumes that close to the wall u varies linearly with y, that is,

$$u = \lambda y. \tag{P4.25}$$

Eq. (P4.24) with u given by Eq. (P4.25) can be transformed to an ordinary differential equation by introducing a new variable ξ, where

$$\xi = y\left(\frac{\lambda \Pr}{9\nu x}\right)^{1/3}. \tag{P4.26}$$

Then the resulting equation becomes

$$\frac{d^2 T}{d\xi^2} + 3\xi^2 \frac{dT}{d\xi} = 0. \tag{P4.27}$$

For the boundary conditions given by

$$\xi = 0, \quad T = T_w : \xi \to \infty; \quad T \to T_e, \tag{P4.28}$$

Eq. (P4.27) has the following solution

$$\frac{T - T_w}{T_e - T_w} = \frac{1}{0.893}\int_0^\xi e^{-\xi^3}\, d\xi. \tag{P4.29}$$

(a) Show that according to the Leveque solution the local Nusselt number Nu_x defined by Eq. (4.4.24a) is

4. Laminar Boundary Layers

$$\text{Nu}_x = \frac{x}{0.893}\left(\frac{\lambda \text{Pr}}{9\nu x}\right)^{1/3}. \tag{P4.30}$$

(b) Using the Blasius solution to evaluate λ, obtain an expression for Nu_x and compare its accuracy with the exact solutions for small, moderate and high Prandtl number fluids. Discuss the results.

SOLUTION

a. With ξ given by (P4.26) and for a dimensionless temperature profile given by (P4.29), we can write

$$\left(\frac{\partial T}{\partial y}\right)_w = (T_e - T_w)\frac{\partial q}{\partial \xi}\frac{\partial \xi}{\partial y} = \frac{T_e - T_w}{0.893}\left(\frac{\lambda \text{Pr}}{9\nu x}\right)^{1/3}$$

$$\therefore \ \text{Nu}_x = -\frac{x}{T_w - T_e}\left(\frac{\partial T}{\partial y}\right)_w = \frac{x}{0.893}\left(\frac{\lambda \text{Pr}}{9\nu x}\right)^{1/3}.$$

b. For Blasius flow

$$\lambda = \left(\frac{\partial u}{\partial y}\right)_w = 0.332\frac{u_e}{x}\sqrt{R_x}$$

$$\text{Nu}_x = \frac{x}{0.893}\left(\frac{0.332 u_e \text{Pr}}{9\nu x^2}\sqrt{R_x}\right)^{1/3} = 0.373 R_x^{1/2}\text{Pr}^{1/3},$$

or

$$\text{Nu}_x R_x^{-1/2} = 0.373\text{Pr}^{1/3}.$$

Comparison with exact solution

Pr	(P4.30)	Exact. Sol.	% Error
0.1	0.173	0.139	25%
1	0.373	0.332	12%
10	0.804	0.729	10%
$\to \infty$	$0.373\text{Pr}^{1/3}$	$0.339\text{Pr}^{1/3}$	10%

As can be seen from the above table, the error by the Leveque approximation decreases as the Prandtl number increases with an asymptotic value of 10% resulting from the neglect of the normal component convective term in the energy equation whose solution is obtained with the Leveque approximation.

4.16 PROBLEM

For a two-dimensional flow near a stagnation point ($u_e = Cx$), calculate δ^*, θ, τ_w by Thwaites' method and $\dot{q}_w$ by the Smith-Spalding method. Compare the results with the exact solutions.

SOLUTION

In (4.5.21a) the second term on the RHS is 0 at the stagnation point since $u_{e_i} = 0$, and $u_e = cx$ for flow near the stagnation point,

$$\therefore \quad \theta^2 = \frac{0.45\nu}{(cx)^6} \int_0^x c^5 x^5 dx \quad \text{or} \quad = 0.075 x^2 / R_x,$$

or

$$\frac{\theta}{x} = 0.27386 R_x^{-1/2}.$$

With

$$\lambda \equiv \frac{\theta^2}{\nu} \frac{du_e}{dx} = \frac{0.075 x^2 R_x^{-1}}{\nu} \frac{u_e}{x} = 0.075,$$

it follows from (4.5.23) that

$$H = 2.61 - 3.75\lambda + 5.24\lambda^2 = 2.3582,$$

$$\delta^* = H\theta = 0.64582 \times R_x^{-1/2}, \quad \frac{\delta^*}{x} = 0.64582 R_x^{-1/2},$$

$$\frac{c_f}{2} = \frac{f_w''}{\sqrt{R_x}} = R_\theta^{-1}(0.225 + 1.61\lambda - 3.75\lambda^2 + 5.24\lambda^3),$$

$$= R_x^{-1/2} \frac{0.225 + 1.61 \times 0.075 - 3.75 \times 0.075^2 + 5.24 \times 0.075^3}{0.2783},$$

$$\therefore \quad f_w'' = \sqrt{R_x} \frac{c_f}{2} = 1.19355.$$

According to the Smith-Spalding method, St is given by (4.5.31) which, for $u_e = cx$ and with $c_2 = (c_3 - 1)/2$, can be expressed as

$$\text{St} = \frac{c_1 (u_e^*)^{c_2}}{[c^{c_3}/(1+c_3)(x^*)^{c_3+1}]^{1/2}} R_L^{-1/2}$$

$$= c_1 (1 + c_3)^{1/2} (R_L u_e^* x^*)^{-1/2}$$

$$= c_1 (1 + c_3)^{1/2} R_x^{-1/2}.$$

for Pr = 1, $c_1 = 0.332$ and $c_3 = 1.95$, St $= 0.5702 R_x^{-1/2}$,
and $g_w' = \text{Nu}_x/\sqrt{R_x} = \text{St} \Pr R_x/\sqrt{R_x} = 0.5702$.
The approximate and exact solutions for Pr = 1 are shown below and indicate good agreement

	Approx. Sol.	Exact Solution (Tables 4.1 and 4.3)
f_w''	1.19355	1.23259
δ_1^*	0.64582	0.64791
θ_1	0.27386	0.29234
g_w	0.5702	0.5708

4.17 Problem

Repeat Problem 4.16 for a flat plate.

Solution

For the Blasius flow, since $u_e = $ const, and $\lambda = 0$, it follows from Thwaites' method that

$$\left(\frac{\theta}{L}\right)^2 R_L = \frac{0.45}{(u_e^*)^6} \int_0^{x^*} (u_e^*)^5 dx^* = 0.45 x^*$$

or

$$\frac{\theta}{x} = 0.6708 R_x^{-1/2} \rightarrow \theta_1 = 0.6708$$

and

$$H = 2.61, \quad \frac{\delta^*}{x} = 1.7508 R_x^{-1/2} \rightarrow \delta_1^* = 1.7508, \quad f_w'' = \frac{c_f}{2}\sqrt{R_x} = 0.335.$$

According to the Smith-Splading method, St is given by (4.5.31) which, with $c_3 = 2c_2 + 1$, can be written as

$$\text{St} = \frac{c_1 (u_e^*)^{c_2}}{[(u_e^*)^{2c_2+1} x^*]^{1/2}} R_L^{-1/2} = c_1 R_x^{-1/2}.$$

For $\Pr = 1.0$ and $c_1 = 0.332$, $\text{St} = 0.332 R_x^{-1/2}$,

$$g_w' = \frac{\text{Nu}_x}{\sqrt{R_x}} = \frac{\text{St} \Pr R_x}{R_x^{1/2}} = 0.332.$$

The approximate and exact solutions are shown below and are in excellent agreement.

	f_w''	δ_1^*	θ_1	g_w'
Thwaites or Smith-Spalding	0.335	1.7508	0.6708	0.332
Exact	0.332	1.7207	0.6641	0.332

4.18 Problem

Repeat Problem 4.16 for an axisymmetric stagnation-point flow ($u_e = Cx^{1/3}$).

Solution

For an axisymmetric stagnation-point flow with $u_e^* = cx^{*1/3}$ it follows from (4.5.21b) that

$$\left(\frac{\theta}{L}\right)^2 = \frac{0.45}{R_L(u_e^*)^6} \int_0^{x^*} c^5 (x^*)^{5/3} dx^* = 0.16875 (x^*)^2/R_x$$

or
$$\theta_1 = \frac{\theta}{x} R_x^{1/2} = 0.41079$$

and
$$\lambda = \frac{\theta^2}{\nu} \frac{du_e}{dx} = \frac{(0.41079 \times R_x^{-1/2})^2}{\nu} \frac{1}{3} \frac{u_e}{x} = 0.05625$$

$$H = 2.61 - 3.75\lambda + 5.24\lambda^2 = 2.4156,$$

$$\delta_1^* = \frac{\delta^*}{x} R_x^{1/2} = 0.99232$$

$$f_w'' = \frac{c_f}{2} R_x^{1/2} = \frac{R_x^{1/2}}{R_\theta}(0.225 + 1.61\lambda - 3.75\lambda^2 + 5.24\lambda^3)$$
$$= 0.30463 R_x^{1/2}/(0.41079 R_x^{1/2}) = 0.74157.$$

Substituting $u_e^* = cx^{*1/3}$ into (4.5.31) and integrating, we get

$$\text{St} = c_1(u_e^*)^{c_2} R_L^{-1/2} \left(c^{c_3} \frac{3}{c_3+3} x^{*(c_3+3)/3} \right)^{-1/2} = c_1 \left(\frac{c_3+3}{3} \right)^{1/2} R_x^{-1/2}.$$

With
$$\text{Pr} = 1.0, \quad \text{St} = 0.42646 R_x^{-1/2}, \quad g_w' = \frac{\text{Nu}_x}{\sqrt{R_x}} = \frac{\text{St Pr} R_x}{\sqrt{R_x}} = 0.42646.$$

Comparison with exact solutions (Pr = 1) indicates good agreement, as shown by the table.

	Approx. Sol.	Exact Solution
f_w''	0.74157	0.75745
δ_1^*	0.99232	0.98536
θ_1	0.41079	0.42900
g_w'	0.42646	0.4402

4.19 Problem

Using Thwaites' method, compute θ/L, δ^*/L and c_f for a flow in which the external velocity varies linearly with distance as

$$\frac{u_e}{u_\infty} = 1 - a\frac{x}{L} \tag{P4.31}$$

as a function of x/L for $a = 1/8$ with $R_L = 10^6$. Find also the location of flow separation. Note that this flow is usually referred to as Howarth's flow. Take $\Delta x = 0.02$. See Section 11.9.

4. Laminar Boundary Layers

SOLUTION

In Thwaites' method, the boundary-layer parameters, θ, H, and c_f are calculated from the formulas given by (4.5.21b) and (4.5.23) Eq. (4.5.21b) was integrated numerically with the trapezoidal rule and $c_f/2$ and H were calculated from (4.5.23) for the Howarth flow, $u_e^* = 1 - ax^*$, for $a = 1/8$ and $R_L = 10^6$. Uniform spacing of $\Delta x^* = 0.02$ was taken between $x^* = 0.0$ and 1.0. The calculated c_f distribution is given below and the computer program is given in the accompanying CD. From the c_f distribution, it is evident that the flow separates between $x^* = 0.98$ and 1.0 where either $\lambda < -0.09$ or $c_f < 0.0$.

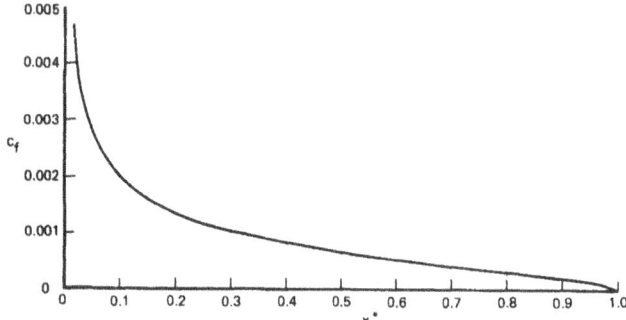

4.20 PROBLEM

Calculate the Stanton number distribution for the external velocity distribution given by Eq. (P4.31) for $\Pr = 1$ and 10 and plot the ratio of $\mathrm{St}/(c_f/2)$.

Hint: See Section 11.9.

SOLUTION

To calculate St, we use (4.5.31) and to calculate $c_f/2$ we use (4.5.21b) and (4.5.23). For $u_e^* = 1.0 - ax^*$, both (4.5.31) and (4.5.21b) were integrated with the trapezoidal rule for a uniform spacing of $\Delta x^* = 0.02$. The calculations were started at $x^* = 0.0$ and were continued up to $x^* = 1.0$. Separation occured between $x^* = 0.98$ and 1.0. The distributions of the Stanton number and the ratio of St to $c_f/2$ for $\Pr = 1.0$ and 10.0 are shown below and indicate that St increases as Pr decreases and that the Reynolds analogy $(\mathrm{St}/(c_f/2) \sim 1.0)$ is valid only for $\Pr = 1.0$ and for $x^* \ll 1.0$ where the pressure gradient is very mild and the flow is far from separation. The computer program is given in the accompanying CD.

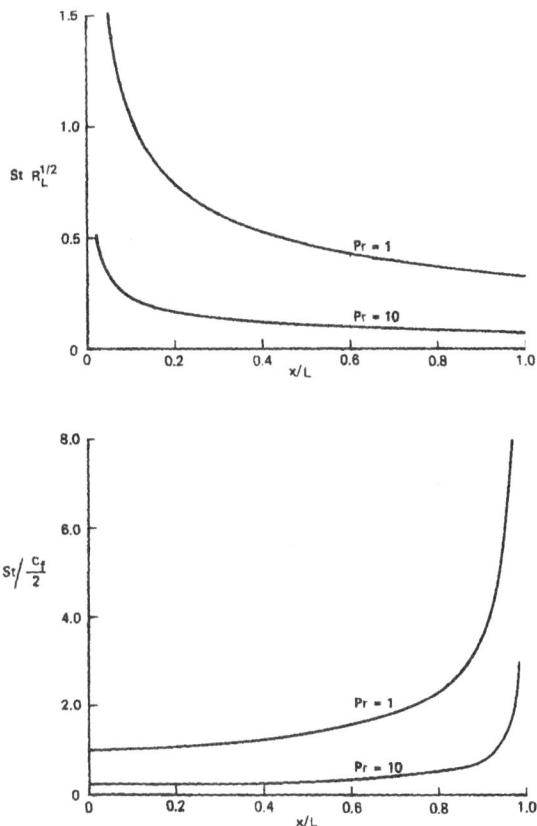

4.21 Problem

Air at 15 °C and at a pressure of 1 atm is flowing over a uniformly heated ellipse of thickness ratio 1/4 and wall temperature of $T_w = 30\,°\mathrm{C}$ at zero angle of attack at a velocity of $10\,\mathrm{m\,s^{-1}}$. Assuming the flow to be laminar up to separation, compute

- (a) Local skin-friction coefficient
- (b) Location of flow separation
- (c) Wall heat flux rate, in $\mathrm{kW\,m^{-2}}$.

Take $\mathrm{Pr} = 0.7$, $a = 1$.

Hint: See Section 11.9, Test Case 3.

Solution

The local skin-friction coefficient, c_f, and wall heat flux rate, $\dot{q}_w$, are calculated by integrating for $u_e^* = 1.25\cos\beta$ with $A = 11.68$ and $B = 2.87$ for $\mathrm{Pr} = 0.7$. The results shown below indicate flow separation at $x/a = 1.66$. The computer program is given in the accompanying CD.

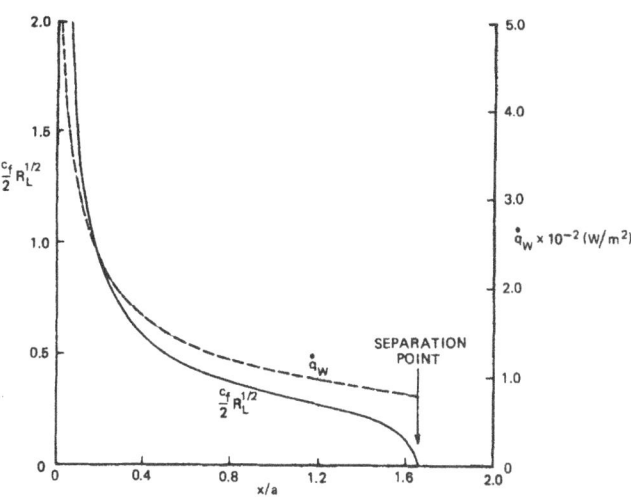

4.22 Problem

Air at 15 °C and at a pressure of 1 atm is flowing over a uniformly heated NACA 0012 airfoil at a wall temperature of 25 °C and at a velocity of $20\,\mathrm{m\,s^{-1}}$. Assuming the flow to be laminar up to separation, with $c = 1\,\mathrm{m}$, compute

(a) Local skin-friction coefficient
(b) Location of flow separation
(c) Wall heat flux rate, in $\mathrm{W\,m^{-2}}$.

The coordinates of the airfoil and its external (inviscid) velocity distribution are given in Table 11.3.1 in the accompanying CD (see Section 11.3).

Solution

The calculated values of c_f and $\dot{q}_w$ are shown in the figure given below. The flow separates at about $x/a = 0.59$ where the pressure gradient is adverse and the Stanton number departs from $c_f/2$ at about $x/a = 0.15$ where the flow begins to decelerate. The computer program is given in the accompanying CD.

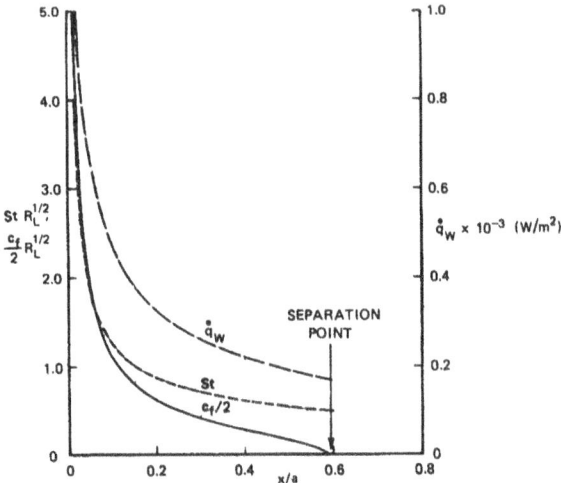

4.23 PROBLEM

A prolate spheroid is an ellipsoid of revolution whose length along its symmetry axis is greater than the diameter of its largest circular cross section. The equation of a prolate spheroid whose center is at $(a, 0)$ can be written, analogous to the equation for an ellipse (see Section 11.9), as

$$\frac{(x-a)^2}{a^2} + \frac{r_0^2}{b^2} = 1. \tag{P4.32}$$

According to inviscid flow theory, for zero angle of attack, the external velocity distribution around the prolate spheroid is given by

$$u_e(s) = u_\infty A \cos\beta. \tag{P4.33}$$

Here s represents the surface distance and β is given by

$$\beta = \tan^{-1}\frac{dy}{dx}. \tag{P4.34}$$

The parameter A is a function of the thickness ratio t ($\equiv b/a$) of the elliptic profile. It is given by

$$A = \frac{(1-t^2)^{3/2}}{\sqrt{1-t^2} - \frac{1}{2}t^2 \ln[(1+\sqrt{1-t^2})/(1-\sqrt{1-t^2})]}. \tag{P4.35}$$

Show that for a sphere ($t = 1$), Eq. (P4.33) reduces to Eq. (E4.9.1).

SOLUTION

As shown in the figure, $\beta = \tan^{-1}(dr_0/dx)$, $\theta = \pi - (\alpha + \beta)$ and for a sphere, $\alpha = \pi/2$ and $\theta = \pi/2 - \beta$. To show that A, given by (P4.40),

4. Laminar Boundary Layers

approaches 1.5 as $t \to 1.0$, we set $t^2 = 1 - \varepsilon^2$ where $\varepsilon \to 0$ as $t \to 1.0$. Then
$$(1 - t^2)^{3/2} = \varepsilon^3,$$

$$\sqrt{1 - t^2} - \frac{1}{2} t^2 \ln[(1 + \sqrt{1 - t^2})/(1 - \sqrt{1 - t^2})]$$
$$= \varepsilon - \frac{1}{2}(1 - \varepsilon^2) \ln[(1 + \varepsilon)/(1 - \varepsilon)]$$
$$= \varepsilon - \frac{1}{2}(1 - \varepsilon^2) \left[(\varepsilon - \frac{1}{2}\varepsilon^2 + \frac{1}{3}\varepsilon^3 \ldots) \right.$$
$$\left. -(-\varepsilon + \frac{1}{2}\varepsilon^2 - \frac{1}{3}\varepsilon^3 \ldots) \right] = \frac{2}{3}\varepsilon^3 + O(\varepsilon^4)$$

$$\therefore \quad A = \varepsilon^3 / \left[\frac{2}{3}\varepsilon^3 + O(\varepsilon^4) \right] = 1.5 \quad \text{as} \quad \varepsilon \to 0$$

$$u_e = u_\infty A \cos \beta = 1.5 u_\infty \cos \left(\frac{\pi}{2} - \theta \right) = 1.5 u_\infty \sin \theta.$$

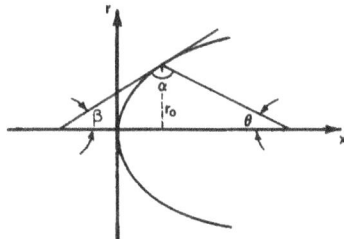

4.24 Problem

Water at 10 °C and at a pressure of 1 atm is flowing over a uniformly heated prolate spheroid of thickness ratio 4 and wall temperature of $T_w = 25\,°\text{C}$ at a velocity of $5\,\text{m\,s}^{-1}$. Assuming the flow to be laminar, with $a = 0.1\,\text{m}$, compute

(a) Local skin-friction coefficient
(b) Location of flow separation
(c) Wall heat flux rate, in $\text{W\,m}^{-2}$.

Solution

The calculated values of c_f, $\dot{q}_w$ and $\text{St}/(c_f/2)$ are plotted in the figure given below. The flow separates at $x/a = 1.62$ where $\lambda = -0.09$. Here x is measured along the major axis from the stagnation point. The computer program is given in the accompanying CD.

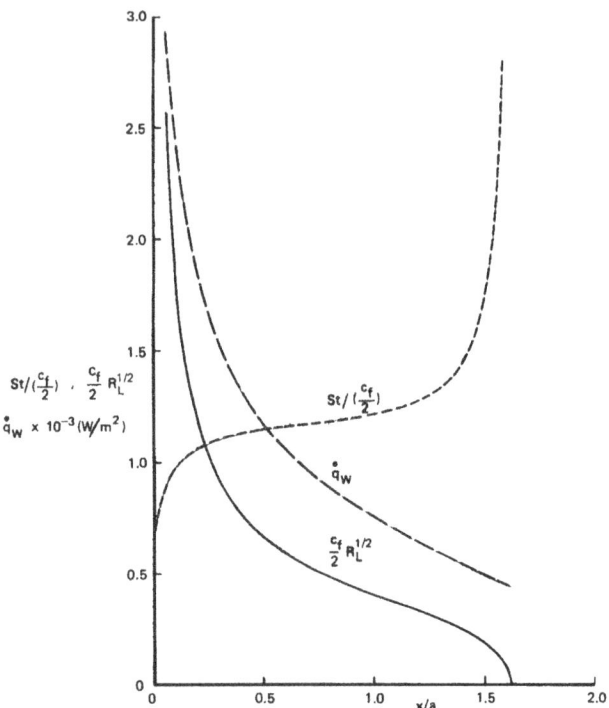

4.25 Problem

Consider the flow situation of Example 4.10 with the adiabatic-wall replaced by a wall with a heat-flux distribution which leads to a uniform wall temperature and the value of δ/y_c changed to 0.95. If the free-stream temperature is 350 °K and the wall temperature 300 °K, calculate the local heat-transfer rate at $(x-x_0)/y_c = 16$ and state, with reasons, whether this value would increase or decrease if the value of δ/y_c was increased to 1.95.

Solution

For $\xi - \xi_0 = 16$, $u_c/u_e = 1.0$ and $\delta/y_c = 0.95$. Then, from Fig. 4.18

$$-g'_w = 0.24 = \frac{+\dot{q}_w}{k(T_w - T_e)\sqrt{\frac{u_e}{\nu x}}}.$$

For $\delta/y_c = 0.95$ and $y_c = 3.0\,\text{mm}$, $\delta = 3.0 \times 0.95 = 2.85\,\text{mm}$, $R_\delta = 190$

$$R_{x_0} = \left(\frac{190}{5.3}\right)^2 = 1285, \quad x_0 = 1.93\,\text{mm}$$

$$\therefore \quad x = 16 y_c + x_0 = 16 \times 0.3 + 1.93 = 4.82\,\text{cm}$$

$$\dot{q}_w = -g'_w k(T_w - T_e)\sqrt{\frac{u_e}{\nu x}} = 0.24 \times 0.026 \times 50$$

$$\times \left(\frac{1.0}{1.5 \times 10^{-5} \times 0.048}\right)^{1/2} = 368\,\text{W/m}^2.$$

The value would decrease if δ/y_c increases to 1.95 in accord with Fig. 4.15 and Reynolds analogy.

4.26 PROBLEM

Compute u_c, δ/h, $\dot{m}/\varrho$ for a two-dimensional laminar jet issuing into still air assuming a half duct width, $h = 5\,\text{cm}$, $\nu = 1.5 \times 10^{-5}\,\text{m}^2\,\text{s}^{-1}$ and duct velocity profile to be:

(a) parabolic; that is,

$$u = u_{\max}[1 - (y/h)^2];$$

(b) uniform

$$u - u_{\max} = \text{const.}$$

In each case, take $u_{\max} = 0.5\,\text{m}\,\text{s}^{-1}$.

SOLUTION

u_c, $\frac{\delta}{h}$ and $\dot{m}/\dot{\varrho}$ for a 2-D laminar jet follow from (8.29a–c) in the book.

a. Parabolic velocity profile: $u = u_{\max}[1 - (y/h)^2]$;

$$\frac{J}{\varrho} = 2\int_0^x u^2\,dy = 2u_{\max}^2 \int_0^h \left[1 - \left(\frac{y}{h}\right)^2\right]^2 dy = \frac{16}{15} u_{\max}^2 h,$$

$$u_c = \left[\frac{3}{32}\left(\frac{16}{15} \times 0.5^2 \times 0.05\right)^2 \frac{1}{1.5 \times 10^{-5} \times 0.05}\right]^{1/3} \left(\frac{h}{x}\right)^{1/3}$$

$$= 2.81 \left(\frac{h}{x}\right)^{1/3}\,\text{m/s},$$

$$\frac{\delta}{h} = \left(\frac{12\sqrt{2}\nu^2 x^2}{16/15 u_{\max}^2 h^4}\right)^{1/3} = \left(\frac{12\sqrt{2}\nu^2}{16/15 \times 0.5^2 \times 0.05^2}\right)^{1/3} \left(\frac{x}{h}\right)^{2/3}$$

$$= 1.79 \times 10^{-2} \left(\frac{x}{h}\right)^{2/3},$$

$$\frac{\dot{m}}{\varrho} = 3.302 \left(\frac{J}{\varrho}\right)^{1/3} (\nu h)^{1/3} \left(\frac{x}{h}\right)^{1/3}\,\text{m}^3/\text{s}$$

$$= 3.302 \left(\frac{16}{15} \times 0.5^2 \times 0.05\right)^{1/3} (1.5 \times 10^{-5} \times 0.05)^{1/3} \left(\frac{x}{h}\right)^{1/3}$$

$$= 7.114 \times 10^{-3} \left(\frac{x}{h}\right)^{1/3}.$$

b. Uniform flow: $u = u_{\max} = $ const :

$$\frac{J}{\varrho} = 2\int_0^\infty u^2 dy = 2u_{\max}^2 h$$

$$u_c = \left[\frac{3}{32} \times (2 \times 0.5^2 \times 0.05)^2 \frac{1}{1.5 \times 10^{-5} \times 0.05}\right]^{1/3} \left(\frac{h}{x}\right)^{1/3}$$

$$= 4.275 \left(\frac{h}{x}\right)^{1/3} \text{ m/s},$$

$$\frac{\delta}{h} = \left[\frac{12\sqrt{2} \times (1.5 \times 10^{-5})^2}{2 \times 0.5^2 \times 0.05^2}\right]^{1/3} \left(\frac{x}{h}\right)^{2/3} = 1.451 \times 10^{-2} \left(\frac{x}{h}\right)^{2/3},$$

$$\frac{\dot{m}}{\varrho} = 3.302(2 \times 0.5^2 \times 0.05 \times 1.5 \times 10^{-5} \times 0.05)^{1/3} \left(\frac{x}{h}\right)^{1/3}$$

$$= 8.77 \times 10^{-3} \left(\frac{x}{h}\right)^{1/3} \text{ m}^3/\text{s}.$$

4.27 PROBLEM

For laminar mixing or two streams of nearly equal velocity ($u_1 - u_2 \ll u_1$), linearize the momentum equation. Discuss the solution, the velocity profile, the relation of the free shear layer to the external flow and the shear stress on the dividing streamline. Compare it with the numerical solutions given in Fig. 4.25 for the cases $u_2/u_1 = 0$ and $u_2/u_1 = 0.5$.

SOLUTION

With

$$u_d = u_1 - u \ll u_1, \quad u\frac{\partial u}{\partial x} = -(u_1 - u_d)\frac{\partial u_d}{\partial x} \approx -u_1\frac{\partial u_d}{\partial x}$$

$$v\frac{\partial u}{\partial y} = -v\frac{\partial u_d}{\partial y} = -u_d\frac{\partial u_d}{\partial x} \ll -u_1\frac{\partial u_d}{\partial x}, \quad \nu\frac{\partial^2 u}{\partial y^2} = -\nu\frac{\partial^2 u_d}{\partial y^2}.$$

With the above approximations, the momentum equations and its b.c. becomes

$$u_1\frac{\partial u_d}{\partial x} = \nu\frac{\partial^2 u_d}{\partial y^2}; \quad y \to \infty: \quad u_d = 0; \quad y \to -\infty: \quad u_d = u_1 - u_2. \quad (1)$$

To obtain a similarity solution, we let $u_d = f(\eta)$, $\eta = y/\delta(x)$.

$$\frac{\partial u_d}{\partial x} = \frac{\partial f}{\partial \eta}\frac{\partial \eta}{\partial x} = -\eta f'\frac{1}{\delta(x)}\frac{d\delta(x)}{dx}, \quad \frac{\partial^2 u_d}{\partial y^2} = f''\frac{1}{\delta^2(x)}. \quad (2)$$

The first eq. in (1) becomes

4. Laminar Boundary Layers

$$-\frac{u_1 \delta}{\nu}\frac{d\delta}{dx}\eta f' = f'' \tag{3}$$

but similarity requires that

$$\frac{u_1 \delta}{\nu}\frac{d\delta}{dx} = c. \tag{4}$$

Take $c = 1/2$ and integrate to get $\delta = (\nu x/u_1)^{1/2}$. Substituting (4) into (3) and integrating twice

$$u_d = c_1 \int_{-\infty}^{\eta} e^{-\eta^2/4} d\eta + c_2. \tag{5}$$

where $c_2 = u_1 - u_2$, $c_1 = -(u_1 - u_2)/\sqrt{4\pi}$. To satisfy b.c. in (1), write

$$u_1 - u = (u_1 - u_2) - \frac{u_1 - u_2}{2\sqrt{\pi}}\int_{-\infty}^{\eta} e^{-\eta^2/4} d\eta$$

$$-\frac{u_1 - u_2}{2} - \frac{u_1 - u_2}{2\sqrt{\pi}}\int_{0}^{\eta} e^{-\eta^2/4} d\eta$$

$$= \frac{u_1 - u_2}{2}\left(1 - \frac{2}{\sqrt{\pi}}\int_{0}^{\eta/2} e^{-\eta^2} d\eta\right)$$

or

$$\frac{u}{u_1} = 1 - \frac{1}{2}\left(1 - \frac{u_2}{u_1}\right)\left[1 - \mathrm{erf}\left(\frac{\eta}{2}\right)\right].$$

Then

$$\tau_0 = \mu\left(\frac{\partial u}{\partial y}\right)_{y=0} = \mu\left(\frac{\partial u}{\partial \eta}\right)\left(\frac{\partial \eta}{\partial y}\right)_{\eta=0}$$

$$= \mu u_1 \left[-\frac{1}{2}\left(1 - \frac{u_2}{u_1}\right)\left(-\frac{2}{\sqrt{\pi}}\right)e^{-\eta^2/4}\right]_{\eta=0}\left(\frac{u_1}{\nu x}\right)^{1/2}$$

or

$$c_f = \frac{\tau_0}{1/2 \varrho u_1^2} = \left(\frac{\nu}{u_1 x}\right)^{1/2}\frac{2}{\sqrt{\pi}}\left(1 - \frac{u_2}{u_1}\right).$$

The present approximate solutions and Lock's exact solutions for $u_2/u_1 = 0$ and 0.501 are shown in the figure. For $u_2/u_1 = 0.5$, the agreement is good, particularly in the upper branch, despite the assumption that $u_d \ll u_1$. For $u_2/u_1 = 0$, the agreement in the upper branch ($\eta > 0$) is still satisfactory whereas in the lower branch, the agreement is poor. In the latter case, the assumption $u_d = u_1 - u \ll u_1$ is obviously violated.

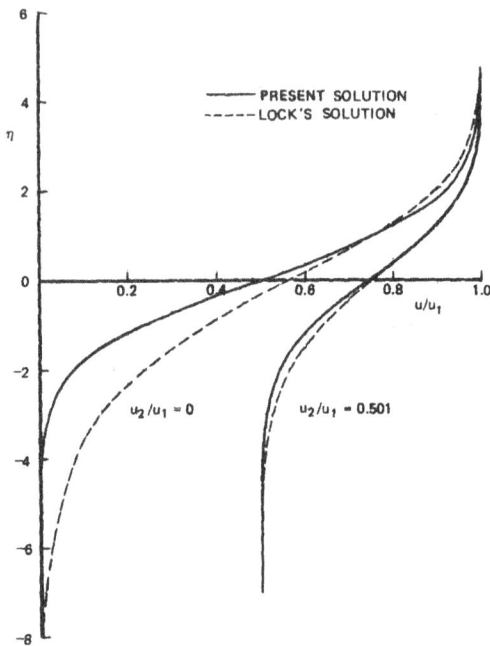

4.28 PROBLEM

Integrate Eq. (4.8.18) for very small values of Pr (~ 0) and for Pr $= 0.5$ and obtain expressions for tine difference between the centerline and the ambient fluid temperature of a two-dimensional laminar jet, similar to Eq. (4.8.19b) which was obtained for Pr $= 1.0$.

SOLUTION

a.

$$\lim_{\text{Pr}\to 0} K = \lim_{\text{Pr}\to 0} K 2\sqrt{2}\varrho c_p N \int_0^\infty \frac{d\eta/\sqrt{2}}{\cosh^{2\text{Pr}+2}(\eta/\sqrt{2})}$$

$$= 2\sqrt{2}\varrho c_p N \int_0^\infty \frac{d\eta/\sqrt{2}}{\cosh^2(\eta/\sqrt{2})} = 2\sqrt{2}\varrho c_p N$$

$$\therefore \quad N = \frac{1}{2\sqrt{2}} \frac{k}{\varrho c_p}.$$

b. For Pr $= 0.5$,

$$K = 2/\sqrt{2}\varrho c_p N \int_0^\infty \frac{d\eta/\sqrt{2}}{\cosh^3(\eta/\sqrt{2})}.$$

Let

$$s = e^x$$

then
$$\int_0^\infty \frac{dx}{\cosh^3 x} = 8 \int_0^\infty \frac{s^2 ds}{(s^2+1)^3} = 8 \left[\int_0^\infty \frac{ds}{(s^2+1)^2} - \int_0^\infty \frac{ds}{(s^2+1)^3} \right]$$

Let
$$t = \tan^{-1} s$$
then
$$dt = \frac{ds}{1+s^2}$$
we have
$$8 \int \cos^2 t \, dt - 8 \int \cos^4 t \, dt$$

Noting that
$$\int \cos^n t \, dt = \frac{\cos^{n-1} t \sin t}{n} + \frac{n-1}{n} \int \cos^{n-2} t \, dt$$
we have
$$= 2 \int \cos^2 t \, dt - 2 \cos^3 t \sin t$$
$$= -2 \sin t \cos^3 t + t + \frac{1}{2} \sin 2t + C$$
$$= \frac{s}{s^2+1} - \frac{2s}{(s^2+1)^2} + \tan^{-1} s + C$$
$$= \tan^{-1} e^x + \frac{e^x}{1+e^{2x}} - \frac{2e^x}{(1+e^{2x})^2} + C$$

Therefore
$$\int_0^\infty \frac{dx}{\cosh^3 x} = \left| \tan^{-1} e^x + \frac{e^x}{1+e^{2x}} - \frac{2e^x}{(1+e^{2x})^2} \right|_0^\infty$$
$$= \frac{\pi}{2} - \frac{1}{2} + \frac{2}{(1+1)^2} = \frac{\pi}{2}$$

5. Laminar Duct Flows

5.1 PROBLEM

Show that for a fully developed laminar flow in a circular pipe,

(a) The average velocity u_m is related to the maximum velocity $u_{\max}$ by

$$u_m = \frac{u_{\max}}{2}, \tag{P5.1}$$

(b) The volume flow rate, $Q(\equiv u_0 \pi r_0^2)$ can be written in the form

$$Q = \frac{\pi r_0^4}{8\mu L}(p_i - p_o) \tag{P5.2}$$

which is known as the Hagen-Poiseuille equation.

SOLUTION

a. The velocity profile in a circular pipe is given by (5.2.4) from which it is clear that the maximum velocity occurs at the center line of the pipe, i.e. $r = 0$ and is equal to

$$u_{\max} = \frac{p_i - p_o}{4\mu L} r_0^2.$$

The mean velocity defined by

$$u_m \equiv \frac{\int u \, dA}{A}$$

then becomes

$$u_m = \frac{p_i - p_o}{4\mu L} r_0^2 \int_0^1 2\left(\frac{r}{r_0}\right)\left[1 - \left(\frac{r}{r_0}\right)^2\right] d\left(\frac{r}{r_0}\right)$$

$$= \frac{1}{2} \frac{p_i - p_o}{4\mu L} r_0^2$$

$$\therefore \quad \frac{u_m}{u_{\max}} = \frac{1}{2}.$$

b. Use the definition of the volume flow rate, $Q = \int u\, dA$, with u given by (5.2.4)

$$Q = \frac{p_i - p_0}{4\mu L} r_0^2 \int_0^{r_0} 2\pi r \left[1 - \left(\frac{r}{r_0}\right)^2\right] dr = \frac{\pi r_0^4}{8\mu L}(p_i - p_0).$$

5.2 PROBLEM

Show that for a fully developed laminar flow between two parallel plates, the solution of the momentum equation (5.2.1), subject to the boundary conditions

$$y = 0, \quad u = 0; \quad y = 2h, \quad u = u_0 = \text{const}$$

is

$$u = u_0 \frac{y}{2h} - \frac{h^2}{2\mu}\frac{dp}{dx}\frac{y}{h}\left(2 - \frac{y}{h}\right). \tag{P5.3}$$

SOLUTION

For a two-dimensional flow, $K = 0$ and (5.2.2) reduces to

$$\frac{dp}{dx} = \mu \frac{d^2 u}{dy^2}$$

with the boundary conditions $y = 0$, $u = 0$; $y = 2h$, $u = u_0$. Integrating this equation twice wrt y and evaluating the two integration constants from the boundary conditions, we get (P5.3).

5.3 PROBLEM

The type of flow represented by Eq. (P5.3) is known as Couette flow. When both plates are stationary ($u_0 \equiv 0$), Eq. (P5.3) becomes

$$u = -\frac{h^2}{2\mu}\frac{dp}{dx}\frac{y}{h}\left(2 - \frac{y}{h}\right), \tag{P5.4}$$

the case of plane Poiseuille flow. Thus Eq. (P5.3) comprises the linear velocity distribution of Eq. (P5.4), due to the shear flow, between the two plates with no imposed pressure gradient, together with the quadratic velocity distribution caused by the pressure parameter, $P[\equiv -(h^2/2\mu u_0)(p_0 - p_1)/L]$.

The flow represented by Eq. (P5.4) is similar to that existing in the narrow clearance between a journal bearing and a stationary shaft, where inertial effects are negligible and the oil is forced along the annulus. Show that

(a) the average velocity u_m is related to the maximum velocity $u_{\max}$ by

$$\frac{u_{\max}}{u_m} = \frac{3}{2} \tag{P5.5}$$

(b) and the dimensionless pressure drop can be written as

$$\frac{p_l - p_0}{\varrho u_m^2} = \frac{3}{R_h}\frac{L}{h} \quad \text{where} \quad R_h = \frac{u_m h}{\nu}. \tag{P5.6}$$

(c) Plot u/u_0 against $y/2h$ from Eq. (P5.3) for values of $P = -2, -1, 0, 1, 2$ to illustrate the effect of pressure gradient on the velocity profile.

SOLUTION

a. With u given by (P5.4), it is obvious that the maximum velocity, $u_{\max}$, occurs at $y = h$, $u_{\max} = -(h^2/2\mu)(dp/dx)$. The mean velocity is

$$u_m = \frac{1}{A}\int u\, dA = \int_0^{2h} -\frac{h^2}{2\mu}\frac{dp}{dx}\frac{y}{h}\left(2 - \frac{y}{h}\right)\frac{dy}{2h} = -\frac{h^2}{3\mu}\frac{dp}{dx}$$

$$\therefore \quad \frac{u_{\max}}{u_m} = \frac{3}{2}.$$

b. From part (a),

$$-\frac{dp}{dx} = \frac{3\mu u_m}{h^2}$$

or

$$\frac{p_i - p_0}{L} = \frac{3\mu u_m}{h^2}.$$

Multiply both sides of the equation by $L/\varrho u_m^2$ and set $R_h = u_m h/\nu$

$$\frac{p_i - p_0}{\varrho u_m^2} = \frac{3\nu L}{u_m h^2} = \frac{3L}{hR_h}.$$

c. With

$$P \equiv -\frac{h^2}{2\mu u_0}\frac{dp}{dx} = -\frac{h^2}{2\mu u_0}\frac{p_0 - p_i}{L}$$

substituted into (P5.3), gives

$$\frac{u}{u_0} = \frac{y}{2h} + 4P\frac{y}{2h}\left(1 - \frac{y}{2h}\right).$$

The variation of u/u_0 vs $y/2h$ for the values of $P = -2, -1, 0, 1,$ and 2 is shown in the figure below.

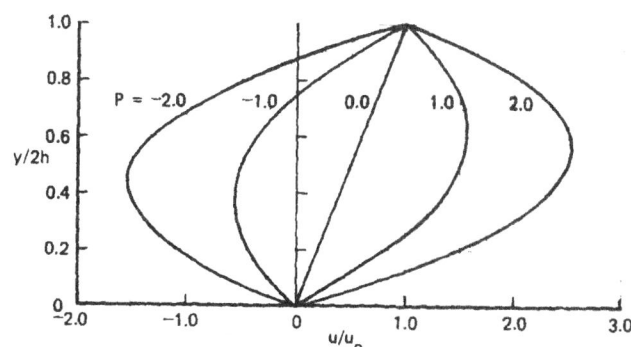

5.4 Problem

Derive Eq. (5.2.23) by following a procedure similar to that used for the constant heat flux rate case. Note that since G is not known prior to the integration, an initial temperature profile must be assumed. Use the temperature profile, Eq. (5.2.16), corresponding to the constant heat flux rate case for this purpose and integrate Eq. (5.1.3) with respect to r. Note that $\partial T/\partial x$ now is given by Eq. (5.2.22). Once a temperature profile is obtained in this way, integrate the energy equation to obtain a new profile. For each temperature profile obtain a mixed mean temperature and a Nusselt number. Repeat this procedure until the successive values of Nusselt number are such that

$$|\text{Nu}^{(i+1)} - \text{Nu}(i)| < \varepsilon$$

where ε is a tolerance parameter, say 10^{-4}.

Solution

For specified wall temperature, the energy equation for the thermally fully-developed pipe flow is given by

$$uG\frac{dT_m}{dx} = \frac{\nu}{\text{Pr}}\frac{1}{r}\frac{\partial}{\partial r}\left(r\frac{\partial T}{\partial y}\right) \tag{1}$$

where $G = (T_w - T)(T_w - T_m)$ is unknown prior to integration. In order to integrate (1), the G profile is initially assumed and updated after a new temperature profile, T, is obtained. For each temperature profile, we compute a Nusselt number and repeat the procedure until successive values of Nusselt number are such that $|\text{Nu}^{(i+1)} - \text{Nu}^{(i)}| \ll \varepsilon$, where ε is a tolerance parameter of, say, 10^{-4}. The computer program for performing the iterative calculations is given in the accompanying CD. After the 7th iteration, the error in Nu is less then 10^{-4} with a converged value of 3.656.

No. of iteration	Nu	Error, ε
1	4.36141	
2	3.72776	0.63366
3	3.66626	0.06149
4	3.65749	0.00877
5	3.65612	0.00138
6	3.65590	0.00022
7	3.65586	0.00004

5.5 Problem

Using the trapezoidal rule, integrate Eq. (5.2.13) numerically for a fully developed thermal boundary layer with constant heat flux rate in the range $0 \leq \bar{r} \leq 1$. Take $\Delta \bar{r} = 0.1, 0.05, 0.025$ and 0.01 and discuss the effect $\Delta \bar{r}$ has on the computed results.

Hint: Express T, u and r in terms of dimensionless quantities. For example, let

$$\bar{u} = \frac{u}{u_m}, \quad \bar{r} = \frac{r}{r_0}, \quad p = \frac{T}{r_0^2 (\mathrm{Pr}/\nu) u_m (dT_m/dx)}$$

and write Eq. (5.2.13) as

$$\frac{d}{d\bar{r}} \left(\bar{r} \frac{dp}{d\bar{r}} \right) = \bar{u} \bar{r}.$$

Integrating from the centerline and setting $dp/d\bar{r} = 0$ at $\bar{r} = 0$, we get

$$\bar{r} \frac{dp}{d\bar{r}} = g.$$

Integrating one more time and taking the mean of the integral, namely, $g_m/\bar{r}_m$, we obtain, with $\bar{p} = p(\bar{r}) - p(0)$,

$$\bar{p} = \int_0^{\bar{r}} \left(\frac{g}{\bar{r}} \right) d\bar{r}.$$

The dimensionless bulk temperature p_m is obtained from

$$p_m = 2 \int_0^1 \bar{r} \bar{u} p \, d\bar{r}$$

and Nusselt number, Nu, from

$$\mathrm{Nu} = \left(\frac{\partial T}{\partial r} \right)_{r=r_0} \frac{d}{T_w - T_m} = \left(\frac{\partial T}{\partial \bar{r}} \right)_{\bar{r}=1} \frac{2}{T_w - T_m} = \frac{2}{\bar{p}_w - \bar{p}_m} \left(\frac{\partial p}{\partial \bar{r}} \right)_{\bar{r}=1}.$$

A listing using the above notation is given below. Note that $\bar{p}_m = p_m - p(0)$ and $\bar{p}_w = \bar{p}(1)$.

```
REAL*8 R(51), U(51), PB(51), G(51)

NP = 51
DO J = 1, NP
    R(J)  = 0.02*DBLE(J-1)
    U(J)  = 2.0 * (1.0 - R(J)**2)
ENDDO

G(1)  = 0.0
PB(1) = 0.0
RU1   = R(1)*U(1)
DO J = 2, NP
    RU2 = R(J) * U(J)
    G(J) = G(J-1) + 0.5*(RU1+RU2)*(R(J) - R(J-1))
    GM   = 0.5*(G(J) + G(J-1))
    RM   = 0.5*(R(J) + R(J-1))
```

```
              PB(J)= PB(J-1) + GM/RM * (R(J) - R(J-1))
              RU1  = RU2
           ENDDO

           T1  = R(1) * U(1) * PB(1)
           PBM = 0.0
           DO J = 2, NP
              T2  = R(J)*U(J)*PB(J)
              PBM = PBM + (T1 + T2) * ( R(J) - R(J-1) )
              T1  = T
           ENDDO
           CNU = 2.0/(PB(NP) - PBM) *G(NP)
           WRITE(6, 100) CNU
           WRITE(6, 110) (J, R(J), PB(J), J=1, NP)

100     FORMAT(1H0, 3X, 'NUSSELT NO = ', E14.6, /)
110     FORMAT(1H0, 3X, 1HJ, 8X, 1HR, 12X, 2HPB/(1H , 2X, I2, 2E14.6))
        END
```

SOLUTION

The calculated Nusselt numbers for $\Delta r = 0.1$, 0.05, 0.025, and 0.01 are given below together with the exact solution.

Δr	Calculated Nu (1)	Exact Nu (2)	err (%) $(1)-(2)/(2)$
0.100	4.3081	4.3636	-1.27
0.050	4.3497	4.3636	-0.32
0.025	4.3602	4.3636	-0.08
0.010	4.3631	4.3636	-0.01

As expected, the error decreases as the step size Δr decreases and the solution approaches its exact value as $\Delta r \to 0$. The listing of the computer program and its output are given in the accompanying CD.

5.6 PROBLEM

Repeat Problem 5.5 for the case of uniform wall temperature.

SOLUTION

For a specified wall temperature, the normalized temperature $G \, (= T_w - T)/(T_w - T_m)$ is unknown prior to integration. The integration procedure described in Prob. 5.4 is used to obtain the solution and the calculated Nusselt numbers for all iterations for step sizes $\Delta r = 0.1$, 0.05, 0.025 and 0.01, are tabulated below. Here ε is defined as $\mathrm{Nu}^{i+1} - \mathrm{Nu}^{(i)}$. Again, the calculated Nusselt number approaches the exact solution $(= 3.658)$ as $\Delta r \to 0$. See the listing and output given in the accompanying CD.

Δr	Calculated Nu (1)	Exact Nu (2)	err (%) (1) − (2)/(2)
0.100	3.5153	3.658	−3.901
0.050	3.5994	3.658	−1.602
0.025	3.6315	3.658	−0.724
0.010	3.6775	3.658	−0.287

5.7 Problem

Show that for a fully developed laminar thermal boundary layer in a two-dimensional channel, the solution of the energy equation is given by

$$\mathrm{Nu} = 7.60 \tag{P5.7a}$$

for the constant surface temperature case and by

$$\mathrm{Nu} = 8.23 \tag{P5.7b}$$

for the constant heat flux rate case.

Solution

a. Constant Heat Flux Rate

The energy equation and its boundary conditions for a 2-D fully-developed channel flow with y measured from the center are:

$$u \frac{dT_m}{dx} = \frac{\nu}{\mathrm{Pr}} \frac{d}{dy}\left(\frac{dT}{dy}\right) \tag{1}$$

$$y = 0, \quad \frac{dT}{dy} = 0; \quad y = \pm h, \quad T = T_w. \tag{2}$$

The velocity profile is $u = 1.5 u_m [1 - (y/h)^2]$ where u_m is the mean velocity and T_m is the mixed mean temperature

$$T_m = \int_0^h uT \, dh / u_m h. \tag{3}$$

Integrate (1) twice wrt y and evaluate the two integration constants by satisfying boundary conditions (2); we get

$$T_w - T = \frac{3}{2} \frac{u_m \mathrm{Pr}}{\nu} h^2 \frac{dT_m}{dx} \left[\frac{5}{12} - \frac{1}{2}\left(\frac{y}{h}\right)^2 + \frac{1}{12}\left(\frac{y}{h}\right)^4\right]. \tag{4}$$

Substituting (4) into (3),

$$T_w - T_m = \frac{3}{2} \frac{u_m \mathrm{Pr}}{\nu} h^2 \frac{dT_m}{dx} \frac{34}{105}. \tag{5}$$

The Nusselt number for a channel flow is defined as

$$\text{Nu} = \frac{\dot{q}_w}{T_w - T_m} \frac{d_e}{k}.$$

Here $d_e \, (= 4h)$ is the hydraulic diameter and

$$\dot{q}_w = -k\left(\frac{\partial T}{\partial y}\right)_w = k\frac{3}{2}\frac{u_m \Pr}{\nu}\frac{dT_m}{dx}h\left(\frac{2}{3}\right). \tag{6}$$

Substituting (5) and (6) into the definition of Nu yields $\text{Nu} = 140/17 = 8.235$.

b. *Uniform Wall Temperature*

In this case, the energy equation is

$$uG = \frac{dT_m}{dx} = \frac{\nu}{\Pr}\frac{d}{dy}\left(\frac{dT}{dy}\right). \tag{7}$$

Intitially we assume $G = 1.0$ which reduces (7) to (1). Equation (7) is then integrated to obtain a temperature profile from which a new G profile is compted to integrate the energy equation again to obtain a new profile. For each temperature profile, a Nusselt number is obtained. Repeat this procedure until sucessive values of Nu are such that $|\text{Nu}^{(i+1)} - \text{Nu}^{(i)}| < \varepsilon$, where $\varepsilon = 10^{-4}$, say. The calculated Nu and the resulting errors with this procedure are listed below.

No. of interation	Nu	Error, (ε)
1	8.23477	
2	7.58131	0.65346
3	7.54422	0.03710
4	7.54126	0.00295
5	7.54101	0.00025
6	7.54099	0.00002

After the 6th interation, the error reduces to 2×10^{-5} and Nu becomes 7.541. These results were obtained for a uniform Δy distribution ($\Delta y = 0.02$) corresponding to 51 point from the centerline to the wall. The computer program and its output are given in the accompanying CD.

5.8 Problem

For liquid metals (very low Prandtl number) the thermal boundary layers fill the tube near the entrance while the velocity distribution is still very nearly uniform (slug flow). An approximation for this case can be obtained by solving Eq. (5.1.3) with u as a constant. Show that under these conditions for a laminar flow in a circular pipe with constant heat flux rate,

$$\text{Nu} = 8.00.$$

5. Laminar Duct Flows

Note that thermal diffusion along the tube can be important with liquid metals but is ignored in this problem.

SOLUTION

The energy equation and its boundary conditions are

$$u\frac{dT_m}{dx} = \frac{\nu}{\Pr}\frac{1}{r}\frac{\partial}{\partial r}\left(r\frac{\partial T}{\partial r}\right), \qquad (1)$$

$$r = 0, \quad \frac{\partial T}{\partial r} = 0; \quad r = r_0, \quad T = T_w. \qquad (2)$$

Integrate (1) wrt r twice and evaluate the two integration constants from b.c.'s given by (2)

$$T_w - T = \frac{\Pr}{4\nu}u\frac{dT_m}{dx}(r_0^2 - r^2). \qquad (3)$$

Note that u is constant across the tube. For the temperature profile given by (3). the mixed mean temperature and heat flux are,

$$T_m = T_w - \frac{\Pr}{8\nu}ur_0^2\frac{dT_m}{dx}, \quad \dot{q}_w = -k\left(\frac{\partial T}{\partial y}\right)_w = \frac{\Pr}{4\nu}u\frac{dT_m}{dx}(2r_0)k.$$

The Nusselt number is then

$$\mathrm{Nu} = \frac{\dot{q}_w}{T_w - T_m}\frac{2r_0}{k} = \frac{(\Pr/4\nu)\,u(dT_m/dx)2r_0 k}{(\Pr/8\nu)\,ur_0^2(dT_m/dx)}\frac{2r_0}{k} = 8.0.$$

5.9 PROBLEM

Air at atmospheric pressure and initially at 120 °C is heated as it flows through a 1 cm diameter tube at a velocity of 3 m s^{-1}. Calculate the heat transfer per unit length of tube if a constant-heat-flux condition is maintained at the wall and the wall temperature is 5 °C above the air temperature. Assume the flow is fully developed.

SOLUTION

The heat transfer rate per unit area is

$$\dot{q}_w = \mathrm{Nu}(T_w - T_m)\frac{k}{d}.$$

Here $\mathrm{Nu} = 4.364$, $T_w - T_m = 5$ K, $d = 0.01$ m, $k = 0.03365$ W/m K. Then $\dot{q}_w = 4.364 \times 5.0 \times 0.03365/0.01 = 73.4$ W/m^2 and the heat transfer rate per unit length of the tube is

$$Q_w = \pi d \dot{q}_w = 3.14159 \times 0.01 \times 73.4 = 2.3\,\mathrm{W/m}.$$

5.10 Problem

Oil flows at a rate corresponding to a bulk velocity of $1\,\mathrm{m\,s^{-1}}$, and is cooled from $80\,°\mathrm{C}$ to $60\,°\mathrm{C}$, in an $0.02\,\mathrm{m}$-diameter tube whose surface is maintained at $15\,°\mathrm{C}$. How long must the tube be to accomplish this cooling for a fully developed flow? Assume the following properties for oil: $c_p = 2.1\,\mathrm{kJ\,kg^{-1}\,K^{-1}}$, $k = 0.1\,\mathrm{W\,m^{-1}}$, $\nu = 6 \times 10^{-6}\,\mathrm{m\,s^{-1}}$, $\kappa = 6 \times 10^{-5}\,\mathrm{m^2\,s^{-1}}$.

Solution

For a fully-developed flow,

$$T_w - T_m = \frac{1}{\mathrm{Nu}} \frac{\mathrm{Pr}}{\nu} u_m r_0^2 \frac{dT_m}{dx}.$$

When the wall temperature is constant,

$$\frac{d(T_m - T_w)}{T_m - T_w} = -\frac{\mathrm{Nu}}{\mathrm{Pr}\,R} \frac{1}{r_0} dx$$

or

$$\ln(T_{m_2} - T_w) - \ln(T_{m_1} - T_w) = -\frac{\mathrm{Nu}}{\mathrm{Pr}\,R} \frac{x_2 - x_1}{r_0}.$$

For $\mathrm{Nu} = 3.658$, $T_{m_2} = 60\,°\mathrm{C}$, $T_{m_1} = 80\,°\mathrm{C}$, $T_w = 15\,°\mathrm{C}$,

$$\mathrm{Pr} = \frac{\nu}{\kappa} = \frac{6 \times 10^{-6}}{6 \times 10^{-5}} = 0.1, \quad r_0 = 0.01;$$

$$R = \frac{u_m r_0}{\nu} = \frac{1.0 \times 0.01}{6 \times 10^{-6}} = 1.67 \times 10^3.$$

The length required to accomplish the required cooling is then

$$\Delta x = x_2 - x_1 = \frac{1.67 \times 10^3 \times 0.1 \times 0.01}{3.658} \times [\ln(80 - 15) - \ln(60 - 15)]$$

$$= 0.168\,\mathrm{m}$$

5.11 Problem

Aviation fuel flows at a rate of $2\,\mathrm{kg\,h^{-1}}$ in a $2\,\mathrm{m}$ long, $0.05\,\mathrm{m}$-diameter tube and is heated from $10\,°\mathrm{C}$ to $50\,°\mathrm{C}$. Using Fig. 5.4 and assuming that the wall heat flux is constant, calculate and plot the tube surface temperature and fluid bulk temperature as a function of tube length. Assume that the fluid properties are constant and are given by

$$\varrho = 700\,\mathrm{kg\,m^{-3}}, \quad k = 0.3 \times 10^{-3}\,\mathrm{kW\,m^{-1}\,K^{-1}}$$
$$c_p = 2.3\,\mathrm{kJ\,kg^{-1}\,K^{-1}}, \quad \nu = 3 \times 10^{-3}\,\mathrm{m^2\,s^{-1}}$$

Solution

$$c_p \Delta T \dot{m} = 2\pi r_0 l \dot{q}_w, \quad \dot{q}_w = c_p \Delta T \dot{m}/2\pi r_0 l$$

$$T_w - T_m = \frac{\dot{q}_w}{\text{Nu}}\frac{2r}{k} = \frac{c_p \Delta T \dot{m}}{2\pi r_0 l}\frac{2r_0}{\text{Nu} k} = \frac{c_p \Delta T \dot{m}}{\pi l k \text{Nu}_x}.$$

For $\dot{m} = 2.0\,\text{kg/h} = \frac{2}{3600}\,\text{kg/s}$, $\Delta T = 50 - 10 = 40\,\text{K}$, $l = 2\,\text{m}$, $k = 0.3 \times 10^{-3}\,\text{kW/m K}$, $c_p = 2.35\,\text{kJ/kg K}$,

$$T_w - T_m = \text{Nu}_x^{-1}\frac{2.35 \times 40 \times 2/3600}{3.14159 \times 2.0 \times 0.3 \times 10^{-3}} = 27.7(\text{Nu}_x)^{-1}.$$

With Nu_x given by Fig. 5.4, $T_w - T_m$ wrt x is given as:

$\hat{x}\left(=\dfrac{x}{r_0}\dfrac{1}{\text{Pr} R_d}\right)$	x [cm]	Nu_x	$T_w - T_m$ [K]
0	0	∞	0
0.002	0.0027	12.00	2.31
0.004	0.0054	9.93	2.79
0.010	0.0136	7.49	3.70
0.020	0.027	6.14	4.51
0.040	0.054	5.19	5.34
0.100	0.136	4.51	6.14
∞	∞	4.36	6.35

5.12 Problem

Glycerin at a temperature of 20 °C flows in a pipe of 0.005 m in diameter and 0.6 m in length. If the pressure drop is 2 bar and the pipe surface temperature is maintained at 30 °C, and $T_m = 25$ °C find:
(a) hydrodynamic entrance length
(b) thermal entrance length
(c) the local heat transfer rate at $x = 0.2, 0.4, 0.6$ m.

The properties of glycerin are as follows:

$$\varrho = 1300\,\text{kg m}^{-3} \quad \nu = 3 \times 10^{-3}\,\text{m}^2\,\text{s}^{-1}$$
$$c_p = 2.3\,\text{kJ kg}^{-1}\,\text{K}^{-1} \quad k = 0.3 \times 10^{-3}\,\text{kW m}^{-1}\,\text{K}^{-1}.$$

Solution

Assume that the entrance length is much smaller than 0.6 m. Then

$$u_m = -\frac{r_0^2}{8\mu}\frac{dp}{dx} = \frac{(p_i - p_0)r_0^2}{8\mu L}.$$

For $p_i - p_0 = 2\,\text{bars} = 2 \times 10^5\,\text{N/m}^2 = 2 \times 10^5\,\text{kg/m s}^2$,

$$\mu = \varrho\nu = 1300 \times 3 \times 10^{-3}\,\text{kg/m s} = 3.9\,\text{kg/m s},$$

$$L = 0.6\,\text{m}, \quad r_0 = 0.0025\,\text{m},$$

$$u_m = \frac{2 \times 10^5 \times 0.0025^2}{8 \times 3.9 \times 0.6} = 0.0668\,\text{m/s},$$

$$R_d = \frac{u_m d}{\nu} = \frac{0.0668 \times 0.005}{3 \times 10^{-3}} = 0.111$$

a. *Hydrodynamic Entrance Length*

$$l_v/r_0 = 0.20R$$

$$l_v = 0.1 \times R_d \times r_0 = 0.1 \times 0.111 \times 0.0025 = 2.78 \times 10^{-4}\,\text{m}$$
$$= 2.78 \times 10^{-2}\,\text{cm} \ll 0.6\,\text{m}$$

b. *Thermal Entrance Length*

$$\Pr \equiv \frac{\varrho \nu c_p}{k} = \frac{1300 \times 3 \times 10^{-3} \times 2.3}{0.3 \times 10^{-3}} = 3.0 \times 10^4$$

$$\therefore \quad l_t = \Pr l_v = 3.0 \times 10^4 \times 2.78 \times 10^{-4} = 8.3\,\text{m}$$

c. *The Local Heat Tranfer Rate at* $x = 0.2, 0.4, 0.6\,\text{m}$

$$\dot{q}_w = \frac{(T_w - T_m)k\text{Nu}}{d} = \frac{(30.0 - 25.0) \times 0.3 \times 10^{-3}}{0.005}\text{Nu}$$
$$= 0.3\,\text{Nu}\,\text{kW/m}^2$$

x	$\hat{x}$	Nu	$\hat{q}_w\,[\text{kW/m}^2]$
0.2	4×10^{-3}	7.9	2.37
0.4	8×10^{-3}	6.3	1.89
0.6	1.2×10^{-2}	5.3	1.59

Here

$$\hat{x} = \frac{x - x_0}{2r_0} \frac{1}{\Pr R} = 0.02(x - x_0).$$

5.13 PROBLEM

Air at 1 bar and at $260\,°\text{K}$ flows in an 2 cm-diameter, 3 m long pipe at a velocity of $1\,\text{m\,s}^{-1}$. If the last 2 m portion of the pipe is heated by keeping the wall temperature at $310\,°\text{K}$, find the local heat transfer rates at $x = 1.5$ and $3\,\text{m}$. Assume that the air temperature is constant over the first 1 m of the pipe and $T_m = 285\,°\text{C}$.

5. Laminar Duct Flows

SOLUTION

$$\nu = 13.8 \times 10^{-6}\,\text{m}^2/\text{s}, \quad \text{Pr} = 0.712 \text{ for air at } T_m = 285\,\text{K},$$

$$R = \frac{u_m r_0}{\nu} = \frac{1 \times 0.01}{13.8 \times 10^{-6}} = 725,$$

$$l_v = 0.20 r_0 R = 0.20 \times 0.01 \times 725 = 1.45\,\text{m}.$$

The hydrodynamic length is 1.45 m whereas the heat transfer starts from 1 m. Therefore the flow can be treated as hydrodynamically fully developed in calculating heat transfer parameters. The local heat transfer rate is given by

$$\dot{q}_w = \frac{(T_w - T_m)k\text{Nu}}{d} = \frac{(310 - 285) \times 0.02524}{0.02}\text{Nu} = 31.5\text{Nu}\,[\text{W/m}^2].$$

To determine Nu, use Fig. 5.4. The heat flux at $x = 1.5\,\text{m}$ and $3.0\,\text{m}$ is:

x	$\hat{x}\left(=\dfrac{x-x_0}{2r_0}\dfrac{1}{\text{Pr}R}\right)$	Nu	$\dot{q}_w\,[\text{W/m}^2]$
1.5	0.048	4.2	132.3
3.0	0.291	3.658	115.2

5.14 PROBLEM

Consider the same flow as in Problem 5.13 except that the heat transfer to the 2 m section is applied at a uniform rate of $25\,\text{W}\,\text{m}^{-2}$. Find the surface temperature distribution along the heated section.

SOLUTION

$$T_w - T_m = \frac{\dot{q}_w d}{k\text{Nu}} = \frac{25 \times 0.02}{0.02524}\text{Nu}^{-1} = \frac{19.8}{\text{Nu}}.$$

x	$\hat{x}$	Nu	$T_w - T_m$	T_w
1.5	0.048	4.8	4.13	289.13
3.0	0.291	4.364	4.54	289.54

Here assume that $T_m = 285\,\text{K}$ and obtain Nu from Fig. 5.4.

5.15 PROBLEM

Apply the Leveque solution (see Problem 4.15) to a laminar flow in a circular pipe with uniform wall temperature and develop an expression for local and average Nusselt number for the thermal entry region. Compare your results with more accurate solutions and discuss any discrepancies.

SOLUTION

With the Leveque solution of the temperature distribution given by Eq. (P4.29), it follows that

$$\dot{q}_w = -k\frac{\partial T}{\partial y}\bigg|_{y=0} = -k\frac{\partial T}{\partial \xi}\bigg|_{\xi=0}\frac{\partial \xi}{\partial y} = -\frac{k}{0.893}(T_e - T_w)\left(\frac{\lambda \text{Pr}}{9\nu x}\right)^{1/3}$$

where

$$\lambda = \left(\frac{\partial u}{\partial y}\right)_{y=0} = \frac{\tau_w}{\mu} = \frac{f(1/8)\varrho u_m^2}{\mu}.$$

For laminar flow $f = \frac{64}{R_d}$ so $\tau_w = \frac{64}{R_d}\frac{1}{8}\varrho u_m^2$; as a result

$$\lambda = \frac{64}{R_d}\frac{1}{8\mu}\varrho u_m^2 = \frac{64}{\varrho(u_m d/\mu)}\frac{1}{8\mu}\varrho u_m^2 = 8\frac{u_m}{d}$$

$$\therefore \text{Nu} = \frac{hd}{k} = \frac{\dot{q}_w}{T_w - T_e}\frac{d}{k} = -\frac{k}{0.893}(T_e - T_w)\left(\frac{\lambda \text{Pr}}{9\nu x}\right)^{1/3}\frac{d}{k}\frac{1}{T_w - T_e}$$

$$= \frac{d}{0.893}\left(\frac{\lambda \text{Pr}}{9\nu x}\right)^{1/3}$$

with $\lambda = 8u_m/d$, and $x^+ = x/d$

$$\text{Nu} = \frac{1}{0.893}\left(\frac{8}{9}\frac{u_m d}{\nu}\frac{\text{Pr}}{x^+}\right)^{1/3} = \frac{1}{0.893}\left(\frac{8}{9}\frac{\text{Pe}}{x^+}\right)^{1/3} = 1.076\left(\frac{\text{Pe}}{x^+}\right)^{1/3}.$$

Note that with x^+ defined as x/r_0 instead of x/d and Peclet number, Pe as $\text{Pr}R_d$,

$$\text{Nu} = 1.355\left(\frac{\text{Pe}}{x^+}\right)^{1/3}$$

which is identical to the expression given in Test Case 1, subsection 11.8.5. To obtain an expression for the average Nusselt number, we integrate $\text{Nu}(x)$ wrt x:

$$\overline{\text{Nu}} = \frac{1}{l_t}\int_{x=0}^{x=l_t}\text{Nu}(x)dx$$

and represent the thermal entrance length by

$$\frac{l_t}{r_0} = 0.1\,\text{Pr}R_d \quad \text{or} \quad \frac{l_t}{d} = \frac{l_t}{2r_0} = 0.05\,\text{Pe}.$$

Then

$$\overline{\text{Nu}} = \frac{d}{l_t} \int_0^{l_t/d} \text{Nu}(x^+) d\left(\frac{x}{d}\right)$$

$$= \frac{1}{l_t/d} \frac{1}{0.893} \left(\frac{8}{9}\right)^{1/3} \text{Pe}^{1/3} \cdot \frac{(x^+)^{2/3}}{2/3}\Big|_0^{l_t/d}$$

$$= \frac{1}{0.893} \left(\frac{8}{9}\right)^{1/3} \text{Pe}^{1/3} \left(\frac{3}{2}\right) \frac{1}{(l_t/d)^{1/3}}$$

$$= \frac{1}{0.893} \left(\frac{8}{9}\right)^{1/3} \left(\frac{3}{2}\right) \text{Pe}^{1/3} \cdot \frac{1}{(0.05\text{Pe})^{1/3}}$$

$$= \frac{1}{0.893} \left(\frac{8}{9}\right)^{1/3} \left(\frac{3}{2}\right) \left(\frac{1}{0.05}\right)^{1/3} = 4.3766.$$

Note that this value of Nusselt number is very close to the exact value of 4.364, given by Eq. (5.2.20). Although the assumption of linear velocity distribution is not a good one, the end result is in good agreement with the exact value. The comparison between the approximate expression $\text{Nu}(x^+) = 1.355(\text{Pe}/x^+)^{1/3}$ and the numerical results obtained from the computer program of Section 11.8, discussed in the solution of Test Case 1, indicate the predictions of both procedures are very good for $x^+ < 0.001\text{Pe}$.

5.16 PROBLEM

(a) Consider a laminar flow in a circular duct in which the velocity field is developed and the temperature field is developing, but in which the thermal shear layers have not yet merged. Using the definition of dimensionless temperature given by Eq. (4.3.7) and the definition of bulk temperature given by Eq. (5.1.4), show that the dimensionless bulk temperature g_m is

$$g_m = 1 + 2 \int_0^{\hat{y}_e} (1 - \hat{y})\hat{u}(g - 1)\, d\hat{y}. \tag{P5.8}$$

Here $\hat{y}_e$ is the dimensionless thermal boundary-layer thickness.

(b) Show that when the thermal shear layers have merged, the dimensionless bulk temperature is

$$g_m = 2 \int_0^1 (1 - \hat{y})\hat{u}g\, d\hat{y}. \tag{P5.9}$$

(c) Show that when the thermal shear layers are merged, the local Nusselt number defined by Eq. (5.2.18) can be written as

$$\text{Nu} = \frac{2}{g_m} g'_w, \tag{P5.10}$$

where $g'_w = (\partial g/\partial \hat{y})_w$. When the thermal shear layers are not merged

$$g'_w = \left(\frac{\partial g}{\partial \eta}\right)_w \frac{1}{\sqrt{\hat{x}}}. \qquad (P5.11)$$

SOLUTION

From (4.3.7) $T = T_e + T_e(1-g)\phi(x)$. Then $T_m = T_e + T_e(1-g)\phi(x)$.

$$\therefore \quad g_m = 1 + \frac{T_m - T_e}{T_e\phi(x)} = 1 + \frac{1}{T_e\phi(x)}\left[\frac{\int_A \varrho u[T_e + T_e(1-g)\phi(x)]dA}{\varrho_m u_m A} - T_e\right]$$

$$= 1 + 2\int_0^1 \hat{u}(1-g)\hat{r}d\hat{r},$$

where $\hat{u} = u/u_m$, $\hat{r} = r/r_0$, and with $\hat{r} = 1 - \hat{y}$, $d\hat{r} = -d\hat{y}$

$$g_m = 1 + 2\int \hat{u}(1-g)(1-\hat{y})(-d\hat{y}) = 1 + 2\int_0^1 \hat{u}(g-1)(1-\hat{y})d\hat{y}.$$

a. When the thermal layers are not merged, with $\hat{y}_e$ corresponding to the thermal-layer thickness, we have

$$g_e = \frac{T_w - T_e}{T_w - T_e} = 1.0 \quad \text{for} \quad \hat{y} > \hat{y}_e$$

$$g_m = 1 + 2\int_0^1 \hat{u}(g-1)(1-\hat{y})d\hat{y}$$

$$= 1 + 2\int_0^{\hat{y}_e} \hat{u}(g-1)(1-\hat{y})d\hat{y} + 2\int_{\hat{y}_e}^1 \hat{u}(g-1)(1-\hat{y})d\hat{y}$$

$$= 1 + 2\int_0^{\hat{y}_e} \hat{u}(g-1)(1-\hat{y})d\hat{y}.$$

b. When the termal layers are merged

$$g_m = 1 + 2\int_0^1 \hat{u}(g-1)(1-\hat{y})d\hat{y}$$

$$= 1 + 2\int_0^1 \hat{u}g(1-\hat{y})d\hat{y} - \int_0^1 \hat{u}(1-\hat{y})d\hat{y}$$

$$= 1 + 2\int_0^1 \hat{u}g(1-\hat{y})d\hat{y} - 1 = 2\int_0^1 \hat{u}g(1-\hat{y})d\hat{y}.$$

c.

$$\mathrm{Nu} = \frac{\dot{q}_w}{T_w - T_m}\frac{d}{k}.$$

Here

$$\dot{q}_w = -k\left(\frac{\partial T}{\partial y}\right)_w = -\frac{k}{r_0}(T_w - T_e)(-g'_w) = \frac{k}{r_0}(T_w - T_e)g'_w$$

$$g_m = \frac{T_w - T_m}{T_w - T_e} \quad \text{or} \quad T_w - T_m = (T_w - T_e)g_m$$

$$\therefore \quad \text{Nu} = \frac{k}{r_0}(T_w - T_e)\frac{d}{k}\frac{1}{(T_w - T_e)g_m}g'_w = \frac{2g'_w}{g_m}.$$

5.17 PROBLEM

In the case of a flow in a circular duct, mass balance gives

$$u_0 \pi r_0^2 = \int_0^{r_0} 2\pi r u\, dr. \tag{P5.12}$$

Show that Eq. (P5.12) in transformed and primitive variables defined by Eqs. (5.4.1) and (5.4.8) can be written as

$$f(x, \eta_{sp}) = \eta_{sp} \tag{P5.13}$$
$$F(x, \sqrt{R_L}) = \sqrt{R_L}/2 \tag{P5.14}$$

respectively.

SOLUTION

a. In transformed variables, we define

$$d\eta = \left(\frac{u_0}{\nu x}\right)^{1/2} \frac{r}{L} dy,$$

$$\eta_{sp} = \int_0^{r_0} \left(\frac{u_0}{\nu x}\right)^{1/2} \frac{r_0 - y}{L} dy = \left(\frac{u_0}{\nu x}\right)^{1/2} \frac{r_0^2}{2L}, \tag{1}$$

$$u_0 \pi r_0^2 = \int_0^{r_0} 2\pi r u\, dr = \int_0^{\eta_{sp}} 2\pi u \left(\frac{\nu x}{u_0}\right)^{1/2} L\, d\eta$$

$$\therefore \quad \int_0^{\eta_{sp}} \frac{u}{u_0} d\eta = \frac{1}{2}\frac{r_0^2}{L}\left(\frac{u_0}{\nu x}\right)^{1/2} = f(x, \eta_{sp}). \tag{2}$$

From (1) and (2), we then have $f(x, \eta_{sp}) = \eta_{sp}$.

b. In primitive variables, with $R_L = u_0 r_0/\nu$, we define

$$dY = \left(\frac{u_0}{\nu L}\right)^{1/2}\left(\frac{r}{L}\right) dy,$$

$$Y_c = \int_0^{r_0} \left(\frac{u_0}{\nu L}\right)^{1/2} \frac{r_0 - y}{L} dy = \frac{1}{2}\sqrt{R_L},$$

$$u_0 \pi r_0^2 = \int_0^{r_0} 2\pi r u\, dr = \int_0^{y_c} 2\pi u \left(\frac{\nu L}{u_0}\right) L\, dY,$$

$$\therefore \quad F(x, Y_c) = \int_0^{Y_c} \frac{u}{u_0} dY = \frac{1}{2}\left(\frac{u_0}{\nu L}\right)^{1/2}\frac{r_0^2}{L} = \frac{1}{2}\sqrt{R_L}.$$

Note that $r_0 = L$.

5.18 Problem

Show that the centerline distance Y_c is equal to $1/2\sqrt{R_L}$ for a circular duct and to $\sqrt{R_L}$ for a plane duct.

Solution

a. For a circular duct, with $R_L = u_0 r_0 / \nu$,

$$dY = \left(\frac{u_0}{\nu L}\right)^{1/2} \frac{r}{L} dy,$$

$$Y_c = \int_0^{r_0} \left(\frac{u_0}{\nu L}\right)^{1/2} \frac{r_0 - y}{L} dy = \left(\frac{u_0}{\nu L}\right)^{1/2} \frac{r_0^2}{2L} = \frac{1}{2}\sqrt{R_L}.$$

b. For a plane duct, with $R_L = \left(\frac{u_0 h}{\nu}\right)$

$$dY = \left(\frac{u_0}{\nu L}\right)^{1/2} dy,$$

$$Y_c = \int_0^h \left(\frac{u_0}{\nu L}\right)^{1/2} dy = \left(\frac{u_0}{\nu L}\right)^{1/2} h = \sqrt{R_L}.$$

5.19 Problem

Derive Eqs. (5.4.13b).

Solution

Along the centerline, $F = \text{constant}$ or $\partial F / \partial x = 0$, and

$$(bF'')' = \left[(1 - \frac{2y}{\sqrt{R_L}})F''\right]'_{y=y_c} = \left[1 - \frac{2y}{\sqrt{R_L}} F''' - \frac{2}{\sqrt{R_L}} F''\right]_{y=y_c}$$

$$= -\frac{2}{\sqrt{R_L}} F''$$

$$(eg')' = -\frac{1}{\Pr} \frac{2}{\sqrt{R_L}} g'$$

$$\therefore \quad -\frac{2}{\sqrt{R_L}} F_c'' = \frac{dp^*}{dx} + F_c' \frac{dF_c'}{dx},$$

$$F_c'' = -\frac{1}{2}\sqrt{R_L} \left(\frac{dp^*}{dx} + \frac{1}{2}\frac{dF_c'^2}{dx}\right)$$

$$g_c' = -\frac{1}{2}\sqrt{R_L}\Pr F_c' \left[\eta(g_c - 1) + \frac{dg_c}{dx}\right]$$

6
Turbulent Boundary Layers

6.1 PROBLEM

Show that the continuity equation requires that $\overline{u'v'}$ should vary as at least the third power of y in the viscous sublayer, whereas the Van Driest formula for mixing length, Eq. (6.4.12) implies $\overline{u'v'} \sim y^4$ for small y.

SOLUTION

In the viscous sublayer, $u^+ = y^+$ or $u = u_\tau^2/\nu y$. From the continuity equation,

$$v = -\int \frac{\partial u}{\partial x} dy = \frac{u_\tau}{\nu} \frac{du_\tau}{dx} y^2$$

so that $u \cdot v \sim y^3$ and $-\overline{u'v'} \sim \sqrt{\overline{u'^2}}\sqrt{\overline{v'^2}} = O(uv) = O(y^3)$ near the wall. Thus $-\overline{u'v'}$ should vary, at least, as the third power of y. From the Van Driest formula for the mixing length,

$$-\overline{u'v'} = \left\{\kappa y \left[1 - \exp\left(-\frac{y^+}{A^+}\right)\right]\right\}^2 \left(\frac{\partial u}{\partial y}\right)^2.$$

In the viscous sublayer, $\partial u/\partial y \sim u_\tau^2/\nu$, and

$$1 - \exp\left(-\frac{y^+}{A^+}\right) \sim 1.0 - \left(1.0 - \frac{y^+}{A^+} \cdots\right) \sim y$$

so $-\overline{u'v'} \sim y^4$.

6.2 PROBLEM

If the expression for the whole velocity profile, Eq. (6.6.1), is evaluated at $y = \delta$, the profile parameter Π can be related to the local skin-friction coefficient $c_f = 2\tau_w/\varrho u_e^2$ and to boundary-layer thickness δ by

$$\sqrt{\frac{2}{c_f}} \equiv \frac{u_e}{u_\tau} = \frac{1}{\kappa} \ln \frac{\delta u_\tau}{\nu} + c + \frac{2\Pi}{\kappa}. \tag{P6.1}$$

Show that it can also be related to the displacement thickness δ^* and to the momentum thickness θ by

$$\kappa \frac{\delta^* u_e}{\delta u_\tau} = 1 + \Pi \qquad (P6.2)$$

and

$$\kappa^2 \frac{(\delta^* - \theta) u_e^2}{\delta u_\tau^2} = 2 + 2\left[1 + \frac{1}{\pi}\text{Si}(\pi)\right]\Pi + \frac{3}{2}\Pi^2. \qquad (P6.3)$$

Also, show that

$$\frac{H}{H-1} \frac{u_\tau}{\kappa u_e} \equiv \frac{1}{\kappa G} = F(\Pi), \qquad (P6.4)$$

where $\text{Si}(\pi) = \int_0^\pi [\sin u/u]\, du = 1.8519$ and G is the Clauser shape parameter.

$$G = \int_0^\infty \left(\frac{u - u_e}{u_\tau}\right)^2 d\left(\frac{y}{\Delta}\right),$$

$$\Delta = -\int_0^\infty \left(\frac{u - u_e}{u_\tau}\right) dy. \qquad (P6.5)$$

SOLUTION

For the velocity profile given by

$$\frac{u}{u_\tau} = \frac{1}{\kappa} \ln \frac{y u_\tau}{\nu} + c + \frac{\Pi}{\kappa}\left[1 - \cos\left(\pi \frac{y}{\delta}\right)\right], \qquad (1)$$

$$\delta^* = \int_0^\delta \left(1 - \frac{u}{u_e}\right) dy = \frac{\delta u_\tau}{u_e} \int_0^\delta \frac{u_e - u}{u_\tau} d\left(\frac{y}{\delta}\right)$$

$$= \frac{\delta u_\tau}{u_e} \int_{e \to 0}^1 \left[-\frac{1}{\kappa}\ln\eta + \frac{\Pi}{\kappa}(1 + \cos \pi\eta)\right] d\eta = \frac{\delta u_\tau}{\kappa u_e}(1 + \Pi) \qquad (2)$$

$$\theta = \int_0^\delta \frac{u}{u_e}\left(1 - \frac{u}{u_e}\right) dy = \frac{\delta u_\tau^2}{u_e^2} \int_0^\delta \left[\frac{u_e}{u_\tau}\left(\frac{u_e - u}{u_\tau}\right) - \left(\frac{u_e - u}{u_\tau}\right)^2\right] d\left(\frac{y}{\delta}\right).$$

After some rearranging, we have

$$\kappa^2 \frac{(\delta^* - \theta) u_e^2}{\delta u_\tau^2} = 2 + 2\left[1 + \frac{1}{\pi}\text{Si}(\pi)\right]\Pi + \frac{3}{2}\Pi^2, \qquad (3)$$

where $\text{Si}(\pi) = \int_0^\pi \frac{\sin u}{u}\, du$. Divide (2) by (3),

$$\frac{\delta^* u_\tau}{\kappa(\delta^* - \theta) u_e} = \frac{1 + \Pi}{2 + 2[1 + (1/\pi)\text{Si}(\pi)]\Pi + 3/2\Pi^2} \equiv F(\Pi) \qquad (4)$$

and from the definition of $H = \delta^*/\theta$, it follows from (4) that

$$\frac{H u_\tau}{\kappa(H-1) u_e} \equiv \frac{1}{\kappa G} = F(\Pi) \quad \text{where} \quad G = \frac{(H-1) u_e}{H u_\tau}.$$

6.3 Problem

Use Eqs. (P6.1) and (P6.2) to find the skin friction on a flat-plate boundary layer at $R_{\delta^*} = u_e \delta^*/\nu = 20{,}000$ and then use Eq. (P6.4) to calculate $R_\theta = u_e \theta/\nu$. Take $c = 5.0$ and $\kappa = 0.41$.

Solution

For a flat-plate boundary layer with high Reynolds number, say, $R_\theta > 5000$, $\Pi = 0.55$ and from (P6.2),

$$\frac{\delta u_\tau}{\nu} = \frac{\kappa}{1+\Pi} \frac{\delta^* u_e}{\nu} = \frac{0.41}{1+0.55} \times 20.000 = 5290.$$

From (P6.1),

$$\left(\frac{2}{c_f}\right)^{1/2} = \frac{1}{\kappa} \ln \frac{\delta u_\tau}{\nu} + c + \frac{2\Pi}{\kappa}$$

$$= \frac{1}{0.41} \ln 5290 + 5.0 + \frac{2 \times 0.55}{0.41} = 28.59$$

$$\therefore \quad c_f = 2.45 \times 10^{-3}.$$

For $\Pi = 0.55$,

$$F(\Pi) = \frac{1+0.55}{2 + 2(1.0 + 1.8519/\pi) \times 0.55 + 1.5 \times 0.55^2} = 0.369.$$

From (P6.5),

$$\frac{H}{H-1} = \kappa\sqrt{2/c_f}\, F(\Pi) = 0.41 \times 28.59 \times 0.369 = 4.325$$

$$\therefore \quad H = 1.301 \quad \text{and} \quad R_\theta = \frac{R_{\delta^*}}{H} = \frac{20.000}{1.301} = 15.370.$$

6.4 Problem

Liquid ammonia at $-25\,°\mathrm{C}$ flows at $2\,\mathrm{m\,s^{-1}}$ past a 2 m-long flat plate whose surface temperature is maintained at $-10\,°\mathrm{C}$. If transition from laminar to turbulent flow occurs at 4×10^5, calculate the local and average heat transfer coefficient $\hat{h}$ at the end of the plate. Evaluate the fluid properties at the arithmetic-mean film temperature T_f.

Solution

From Appendix B, Table B-2: $\varrho = 663\,\mathrm{kg/m^3}$, $c_p = 4518.7\,\mathrm{J/kg\,K}$, $\nu = 3.82 \times 10^{-7}\,\mathrm{m^2/s}$, $\Pr = 2.07$ for liquid ammonia at $T_f = 1/2(T_w + T_e) = (-10 - 25)/2 = -17.5$. At transition $R_{x_{tr}} = 4 \times 10^5$. Then $4 \times 10^5 = 2x_{tr}/(3.82 \times 10^{-7})$, so that, $x_{tr} = 0.076\,\mathrm{m}$, $x_0 = 0.053\,\mathrm{m}$. The local heat transfer coefficient $\hat{h}$ at the end of the plate is

$$\hat{h} = \frac{\dot{q}_w}{T_w - T_e} = \varrho c_p u_e \mathrm{St} = \varrho c_p u_e (0.0296 R_x^{-0.20})$$

$$= 663 \times 4518.7 \times 2.0 \times 0.0296 \left[\frac{2.0 \times (2.0 - 0.053)}{3.82 \times 10^{-7}} \right]^{-0.20}$$

$$= 7033.7 \, \mathrm{W/m^2 \, K}$$

and the average heat transfer coefficient is

$$\hat{h}_m = \int_0^l \hat{h}\, dx/l = \varrho \nu c_p \left(\int_0^{R_{x_{tr}}} \mathrm{St}\, dR_x + \int_{R_{x_{tr}}}^{R_l} \mathrm{St}\, dR_x \right)/l$$

$$= \frac{\varrho \nu c_p}{l} \{ 0.664 \mathrm{Pr}^{-2/3} R_{x_{tr}}^{1/2} \cdot 0.037[(R_l - R_{x_{tr}})^{0.80} - (R_{x_{tr}} - R_{x_0})^{0.80}] \}$$

$$= 663 \times 4518.7 \times 3.82 \times 10^{-7} \{ 0.664 \times 2.07^{-2/3} (4 \times 10^5)^{1/2}$$

$$+ 0.037[(1.047 \times 10^7 - 2.8 \times 10^5)^{0.8}$$

$$- (4 \times 10^5 - 2.8 \times 10^5)^{0.80}]\}/2.0$$

$$= 663 \times 4518.7 \times 3.82 \times 10^{-7} (258.6 + 14526.7)/2.0$$

$$= 8460.39 \, \mathrm{W/m^2 \, K}.$$

6.5 Problem

Using the momentum integral equation (3.5.14) and Eq. (6.7.10), derive Eqs. (6.7.14), (6.7.15), (6.7.12) and (6.7.13). Represent the local skin-friction in Eq. (3.5.14) by

$$c_f = \frac{0.0456}{(u_e \delta/\nu)^{1/4}}$$

which is a modified version of the Blasius friction equation (7.2.16) for friction factors in circular pipes.

Solution

For a power law velocity profile, $u/u_e = (y/\delta)^{1/n}$,

$$\frac{\theta}{\delta} = \frac{n}{(n+1)(n+2)}, \quad \frac{\delta^*}{\delta} = \frac{1}{1+n}. \quad \text{For } n = 7, \quad \frac{\theta}{\delta} = \frac{7}{12}. \quad (1)$$

For a flat plate, $d\theta/dx = c_f/2$ can be written as

$$\frac{7}{72} \frac{dR_\delta}{dR_x} = \frac{c_f}{2} \quad \text{using (1)}. \quad (2)$$

For $c_f = 0.0456 R_\delta^{-1/4}$, it follows from (2), that

$$R_\delta^{1/4} dR_\delta = \frac{0.0456}{2} \frac{72}{7} dR_x, \quad R_\delta = 0.37 R_x^{0.8}$$

or $\delta/x = 0.37 R_x^{-0.2}$, $\theta/x = 0.036 R_x^{-0.2}$,

$$c_f = 0.0456(0.37 R_x^{0.8})^{-1/4} = 0.059 R_x^{-0.2}$$

and

$$\bar{c}_f = \frac{1}{x}\left(\int_0^x c_f\, dx\right) = 0.074 R_x^{-0.2}.$$

6.6 Problem

Consider Example 6.3 where the plate surface is (a) covered with camouflage paint with equivalent sand roughness $k_s = 1 \times 10^{-3}$ cm and (b) cast iron with $k_s = 25 \times 10^{-3}$ cm.

Calculate the momentum thickness, boundary-layer thickness, local skin-friction coefficient and average-skin friction coefficient at $x = 3$ m.

Hint: In calculating θ and δ for rough wall boundary layers on a flat plate, remember that

$$\frac{\bar{c}_f}{2} = \frac{\theta}{x}$$

and a power-law assumption for the velocity profile is a good approximation for δ^*, δ and θ.

Solution

a. Camouflage paint with $k_s = 1 \times 10^{-3}$ cm,

$$\frac{x}{k_s} = \frac{300}{1 \times 10^{-3}} = 3 \times 10^5, \quad R_x = 3 \times 10^7, \quad R_k = 100$$

From Figs. 6.12 and 6.13, c_f and $\bar{c}_f$ are, respectively,

$$c_f = 2.1 \times 10^{-3} \quad \text{and} \quad \bar{c}_f = 2.5 \times 10^{-3}.$$

From the momentum integral equation for a flat plate $\bar{c}_f/2 = \theta/x$ we can write $x(\bar{c}_f/2) = 300(2.5 \times 10^{-3})/2 = 0.375$ cm and $\delta = \frac{72}{7}\theta = 3.857$ cm.

b. For the case of iron with $k_s = 25 \times 10^{-3}$ cm.

$$\frac{x}{k_s} = \frac{300}{25 \times 10^{-3}} = 1.2 \times 10^4, \quad R_x = 3 \times 10^7, \quad R_k = 2500.$$

From Fig. 6.10 and 6.11, c_f and $\bar{c}_f$ are, respectively,

$$c_f = 3.5 \times 10^{-3} \quad \text{and} \quad \bar{c}_f = 4.2 \times 10^{-3}.$$

The momentum thickness and boundary-layer thickness are

$$\theta = \bar{c}_f/2 \cdot x = 0.63 \,\text{cm}, \quad \delta = \frac{72}{7}\theta = 6.48 \,\text{cm}.$$

6.7 Problem

Using the power-law profile for velocity, derive Eq. (6.7.27b). *Hint:* First show that the momentum integral equation for a flat plate can be written as

$$-\frac{\partial}{\partial x}\int_0^y u(u_e - u)\, dy + (u_e - u)\frac{\partial}{\partial x}\int_0^y u\, dy = \frac{\tau - \tau_w}{\varrho}.$$

Solution

For a flat plate flow, write the continuity and momentum equations as

$$\frac{\partial u}{\partial x} + \frac{\partial v}{\partial y} = 0, \quad u\frac{\partial u}{\partial x} + v\frac{\partial u}{\partial y} = \frac{1}{\varrho}\frac{\partial \tau}{\partial y}.$$

Multiply continuity by u and add the resulting expressions to the momentum equation, then integrate the resulting equation wrt y.

$$\int_0^y \frac{\partial}{\partial x}\left(\frac{u^2}{u_e^2}\right) dy - \frac{u}{u_e}\int_0^y \frac{\partial}{\partial x}\left(\frac{u}{u_e}\right) dy = \frac{1}{u_e^2}\frac{\tau - \tau_w}{\varrho}. \tag{1}$$

For a power-law velocity profile,

$$\frac{u}{u_e} = \left(\frac{y}{\delta}\right)^{1/n} = \eta^{1/n} = g(\eta), \quad \eta = \frac{y}{\delta},$$

then

$$\int_0^y \frac{\partial}{\partial x}\left(\frac{u^2}{u_e^2}\right) dy = -\frac{d\delta}{dx}\left(\eta g^2 - \int_0^\eta g^2\, d\eta\right)$$

$$\int_0^y \frac{\partial}{\partial x}\left(\frac{u}{u_e}\right) dy = -\frac{d\delta}{dx}\left(\eta g - \int_0^\eta g\, d\eta\right). \tag{2}$$

Substitute (2) into (1),

$$\frac{\tau}{\tau_w} = 1 + \frac{2}{c_f}\left[\frac{d\delta}{dx}\left(\int_0^\eta g^2\, d\eta - g\int_0^\eta g\, d\eta\right)\right] \tag{3}$$

and evaluate $d\delta/dx$ from $d\theta/dx = c_f/2$. Since $\delta/\theta = $ constant for a power-law profile,

$$\frac{d\delta}{dx} = \frac{d\theta}{dx}\left(\frac{\theta}{\delta}\right)^{-1} = \frac{c_f}{2}\left(\frac{\theta}{\delta}\right)^{-1} \quad \text{or}$$

$$\frac{2}{c_f}\frac{d\delta}{dx} = \left(\frac{\theta}{\delta}\right)^{-1} = \frac{(n+1)(n+2)}{n} \tag{4}$$

then

$$\int_0^\eta g^2\, d\eta = \int_0^\eta \eta^{2/n}\, d\eta = \frac{n}{n+2}\eta^{(n+2)/n}$$

$$g\int_0^\eta g\, d\eta = \eta^{1/n}\int_0^\eta \eta^{1/n}\, d\eta = \frac{n}{n+1}\eta^{(n+2)/n}. \tag{5}$$

6. Turbulent Boundary Layers

Then substitute (4) and (5) into (3),

$$\frac{\tau}{\tau_w} = 1 + \frac{(n+1)(n+2)}{n}\left(\frac{n}{n+2}\eta^{(n+2)/n} - \frac{n}{n+1}\eta^{(n+2)/n}\right)$$

$$= 1 - \eta^{1+2/n} = 1 + \left(\frac{y}{\delta}\right)^{1+2/n}.$$

6.8 Problem

Using Eqs. (6.7.27) and the definition of eddy viscosity and eddy conductivity, derive Eq. (6.7.28) for $\Pr_t = 1$.

Solution

For power-law velocity and temperature profiles,

$$\frac{u}{u_e} = \left(\frac{y}{\delta}\right)^{1/n}, \qquad \frac{T_w - T}{T_w - T_e} = \left(\frac{y}{\delta_t}\right)^{1/n}$$

$$\frac{\partial u}{\partial y} = \frac{u_e}{n}\frac{1}{\delta}\left(\frac{y}{\delta}\right)^{1/n-1}, \qquad -\frac{\partial T}{\partial y} = \frac{T_w - T_e}{n}\left(\frac{y}{\delta_t}\right)^{1/n-1}\frac{1}{\delta_t}.$$

With

$$\Pr_t = 1, \qquad \frac{\varepsilon_m}{\varepsilon_h} = 1.0 = \frac{(\tau/\varrho)/(\partial u/\partial y)}{-\overline{T'v'}/(\partial T/\partial y)}$$

$$\therefore \quad -\overline{T'v'} = \frac{\tau}{\varrho}\frac{\partial T}{\partial y}\left(\frac{\partial u}{\partial y}\right)^{-1} = -\frac{\tau}{\varrho}\frac{T_w - T_e}{u_e}\left(\frac{\delta}{\delta_t}\right)^{1/n}$$

$$\dot{q}_w = \dot{q} = \varrho c_p \overline{T'v'} = \varrho c_p \frac{\tau_w}{\varrho}\frac{T_w - T_e}{u_e}\left(\frac{\delta}{\delta_t}\right)^{1/n}.$$

Then

$$\mathrm{St} = \frac{\dot{q}_w}{\varrho c_p u_e(T_w - T_e)} = \frac{c_p \tau_w (T_w - T_e)}{\varrho c_p u_e^2 (T_w - T_e)}\left(\frac{\delta}{\delta_t}\right)^{1/n} = \frac{c_f}{2}\left(\frac{\delta_t}{\delta}\right)^{-1/n}.$$

6.9 Problem

Air at $u_e/\nu = 3 \times 10^6$ m^{-1} flows past a 3 m-long flat plate. Consider the plate: (a) heated at uniform wall temperature T_w, and (b) the heated portion preceded by an unheated portion x_0 of 1 m. Calculate the Stanton number distribution along the plate for both cases. What role does the term $(T_w/T_e)^{0.4}$ in Eq. (6.7.35) play in the results. Assume the flow to be turbulent from the leading edge with $T_w/T_e = 1.1$ and $\Pr = 0.7$.

SOLUTION

a. From (6.7.35), for $x_0 = 0$,

$$\text{St Pr}^{0.4}\left(\frac{T_w}{T_e}\right)^{0.4} = 0.0296 R_x^{-0.20} = 0.0296 R_L^{-0.20}\left(\frac{x}{L}\right)^{-0.20}.$$

When $R_L = 9 \times 10^5$, $\text{Pr} = 0.7$, $T_w/T_e = 1.1$ and $L = 3\,\text{m}$

$$\text{St} = \frac{0.0296(9 \times 10^5)^{-0.20}}{0.7^{0.4} \times 1.1^{0.4}}\left(\frac{x}{3}\right)^{-0.20} = 2.64 \times 10^{-3} x^{-0.20}. \quad \text{(a)}$$

b. For $x_0 > 0$, (6.7.35) becomes

$$\text{St Pr}^{0.4}\left(\frac{T_w}{T_e}\right)^{0.4} = 0.0296 R_x^{-0.20}\left[1 - \left(\frac{x_0}{x}\right)^{9/10}\right]^{-1/9}.$$

When $R_L = 9 \times 10^5$, $\text{Pr} = 0.7$, $T_w/T_e = 1.1$, $L = 3\,\text{m}$, $x_0 = 1\,\text{m}$,

$$\text{St} = 2.64 \times 10^{-3} x^{-0.20}\left[1 - \left(\frac{1}{x}\right)^{9/10}\right]^{-1/9}. \quad \text{(b)}$$

The variation of St with x for (a) and (b) are shown in the figure

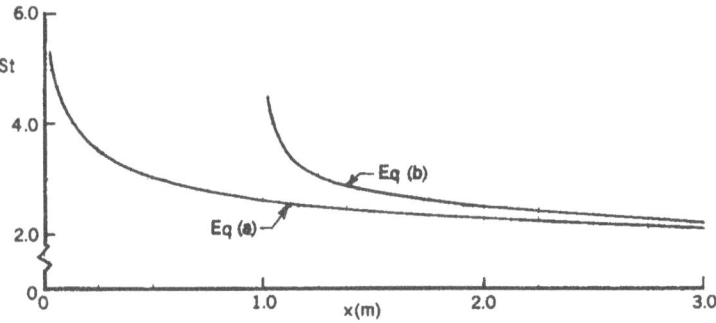

Discussion: Since $(T_w/T_e)^{0.4} > 1.0$, if $T_w/T_e > 1.0$ and vice versa, St decreases as the wall is heated and increases as the wall is cooled.

6.10 PROBLEM

Use Eq. (6.7.38) to derive an expression for wall heat flux on a flat plate for which the difference between wall temperature and freestream temperature varies linearly with x, that is,

$$T_w - T_e = A + Bx.$$

Hint: Note that there is a temperature jump at the leading edge of the plate where $T_w - T_e = A$.

6. Turbulent Boundary Layers

SOLUTION

To calculate the heat flux, $\dot{q}_w$, we use (6.7.38). For the present case

$$\frac{dT_w}{dx} = \frac{d}{dx}(A + Bx) = B, \quad N = 1, \quad x_{01} = 0$$

$$T_w(x_{01}^+) - T_w(x_{01}^-) = T_w(0^+) - T_w(0^-) = A$$

$$\hat{h}(x_{01}, x) = \varrho c_p u_e \text{St}_T \left[1 - \left(\frac{0}{x}\right)^{9/10}\right]^{-1/9} = \varrho c_p u_e \text{St}_T,$$

so

$$\dot{q}_w = \varrho u_e c_p \text{St}_T(x) \left\{ A + B \int_0^x \left[1 - \left(\frac{\xi}{x}\right)^{9/10}\right]^{-1/9} d\xi \right\}$$

$$= \varrho u_e c_p \text{St}_T(x) \left[A + Bx \int_0^1 \left(1 - \eta^{9/10}\right)^{-1/9} d\eta\right]$$

$$\int_0^1 \left(1 - \eta^{9/10}\right)^{-1/9} d\eta = \frac{10}{9} \int_0^1 \xi^{-1/9}(1 - \xi)^{1/9} d\xi = \frac{10}{9} B_1 \left(\frac{8}{9}, \frac{10}{9}\right).$$

Here

$$B_r(m, n) = \int_0^r \xi^{m-1}(1 - \xi)^{n-1} d\xi$$

$$\therefore \quad \dot{q}_w(x) = \varrho u_e c_p \text{St}_T(x) \left[A + \frac{108}{9} \times B_1 \left(\frac{8}{9}, \frac{10}{9}\right)\right].$$

6.11 PROBLEM

Use Eq. (6.7.38) to obtain an expression for the heat transfer rate on a flat plate for $x > x_2$ and with $T_w = T_{w_1}$, for $0 < x < x_1$, $T_w = T_{w_2}$, for $x_1 < x < x_2$, and $T_w = T_{w_3}$, for $x > x_2$.

SOLUTION

The heat flux on a flat plate subject to a series of discontinuous wall temperature is

$$\dot{q}_w = \sum_{n=1}^{N} \hat{h}(x_{\text{on}}, x)[T_w(x_{\text{on}}^+) - T_w(x_{\text{on}}^-)].$$

For the present case, $N = 3$ and

$$n=1, \quad \hat{h}(0,x) = \varrho u_e c_p \mathrm{St}_T(x),$$
$$T_w(0^+) - T_w(0^-) = T_{w_1} - T_e,$$
$$n=2 \quad \hat{h}(x_1,x) = \varrho u_e c_p \mathrm{St}_T(x)\left[1-\left(\frac{x_1}{x}\right)^{9/10}\right]^{-1/9},$$
$$T_w(x_1^+) - T_w(x_1^-) = T_{w_2} - T_{w_1},$$
$$n=3 \quad \hat{h}(x_2,x) = \varrho u_e c_p \mathrm{St}_T(x)\left[1-\left(\frac{x_2}{x}\right)^{9/10}\right]^{-1/9},$$
$$T_w(x_2^+) - T_w(x_2^-) = T_{w_3} - T_{w_2},$$

so

$$\dot{q}_w(x) = \varrho u_e c_p \mathrm{St}_T(x)\left\{(T_{w_1}-T_e) + (T_{w_2}-T_{w_1})\left[1-\left(\frac{x_1}{x}\right)^{9/10}\right]^{-1/9}\right.$$
$$\left. + (T_{w_3}-T_{w_2})\left[1-\left(\frac{x_2}{x}\right)^{9/10}\right]^{-1/9}\right\}.$$

6.12 PROBLEM

Air at 20°C and with $u_e/\nu = 5 \times 10^5\,\mathrm{m}^{-1}$ flows over a 4 m long flat plate. Assume that the flow is turbulent at the leading edge and has the following temperature distribution: 0 to 1 m, 50°C; 1 to 1.5 m, 60°C; 1.5 to 4 m, 80°C: Find the local heat flux at $x = 4$ m for Pr = 1.

SOLUTION

Use the expression for $\dot{q}_w$ given in Problem 6.11 and with $T_{w_1} = 50\,°\mathrm{C}$, $T_{w_2} = 60\,°\mathrm{C}$, $T_{w_3} = 80\,°\mathrm{C}$, $T_e = 20\,°\mathrm{C}$, $x_1 = 1\,\mathrm{m}$, $x_2 = 1.5\,\mathrm{m}$, $x = 4\,\mathrm{m}$, $u_e/\nu = 5 \times 10^5/\mathrm{m}$, and $k = 0.0279\,\mathrm{W/m\,K}$,

$$\dot{q}_w(x) = 5 \times 10^5 \times 0.0279 \times 0.0296 \times (2 \times 10^6)^{-0.20}$$
$$\left\{(50-20) + (60-50)\left[1-\left(\frac{1}{4}\right)^{9/10}\right]^{-1/9}\right.$$
$$\left. + (80-60)\left[1-\left(\frac{1.5}{4}\right)^{9/10}\right]^{-1/9}\right\}$$
$$= 1397\,\mathrm{W/m^2}.$$

6.13 PROBLEM

Air at 20°C and with $u_c/\nu = 5 \times 10^5\,\mathrm{m}^{-1}$ flows over a 5 m long flat plate. Assume that the heating starts at $x_0 = 2\,\mathrm{m}$ and is applied uniformly in the range $x_0 < x \leq 5$, and find the wall temperature distribution along the plate. Assume also that the flow is turbulent from the leading edge.

SOLUTION

The wall temperature is calculated from Eq. (6.7.42), which is

$$T_w(x) - T_e = \frac{33.61\dot{q}_w \mathrm{Pr}^{0.4} R_x^{0.2}}{\varrho c_p u_e} \frac{\beta_r(1/9, 10/9)}{\beta_1(1/9, 10/9)}$$

$$= \frac{33.61\dot{q}_w \mathrm{Pr}^{-0.6} R_x^{0.2}}{u_e/\nu k} \frac{\beta_r(1/9, 10/9)}{\beta_1(1/9, 10/9)}.$$

For $u_e/\nu = 5 \times 10^5/\mathrm{m}$, $R_x = 5 \times u_e/\nu = 25 \times 10^5$, $k = 0.02536$, $\mathrm{Pr} = 0.7$, $x_0 = 2\,\mathrm{m}$, $x = 5\,\mathrm{m}$, $r = 1 - (x_0/x)^{9/10} = 0.5616$, and $\dot{q}_w$ in $\mathrm{W/m^2}$

$$T_w(x) - T_e = \frac{3.61 \times 0.7^{-0.6} \times (2.5 \times 10^6)^{0.2}}{5 \times 10^5 \times 0.02535} \times 0.946 \times \dot{q}_w$$

$$= 0.0591 \dot{q}_w.$$

6.14 PROBLEM

Show that the integral in Eq. (6.7.42) can be expressed by Eq. (6.7.43). *Hint:* To put the integral in the form of the Beta function,

$$B_r(m, n) = \int_0^r z^{m-1}(1-z)^{n-1} dz \quad (m, n > 0)$$

first multiply and divide the integral by x (x is treated as a parameter) and let $z = 1 - (\xi/x)^{9/10}$.

SOLUTION

First, with $\eta = \xi/x$ and $\eta_0 = x_0/x$, consider the integrand,

$$\int_{x_0}^x \left[1 - \left(\frac{\xi}{x}\right)^{9/10}\right]^{-8/9} d\xi = x \int_{\eta_0}^1 \left(1 - \eta^{9/10}\right)^{-8/9} d\eta.$$

Let $z = 1 - \eta^{9/10}$, $dz = -\frac{9}{10}(1-z)^{-1/9} d\eta$ and with $\beta_r(m, n)$ denoting incomplete beta function and $r = 1.0 - (x_0/x)^{9/10}$,

$$x \int_{\eta_0}^1 (1 - \eta^{9/10})^{-8/9} d\eta = x \int_0^r z^{-8/9}(1-z)^{1/9} \frac{10}{9} dz = \frac{10x}{9} \beta_r\left(\frac{1}{9}, \frac{10}{9}\right).$$

Then from Eq. (6.7.41),

$$T_w - T_e = \frac{3.42\dot{q}_w}{\Pr^{0.6} R_x^{0.8} k} \int_{x_0}^{x} \left[1 - \left(\frac{\xi}{x}\right)^{9/10}\right]^{-8/9} d\xi$$

$$= \frac{3.42\Pr^{0.4} R_x^{0.2} \dot{q}_w}{(\mu c_p/k)(u_e x \varrho/\mu) k} \frac{10x}{9} \beta_1\left(\frac{1}{9}, \frac{10}{9}\right) \frac{\beta_r(1/9, 10/9)}{\beta_1(1/9, 10/9)}$$

$$= \frac{3.42 \times 10}{9} \times 8.844 \frac{\dot{q}_w \Pr^{0.4} R_x^{0.2}}{\varrho c_p u_e} \frac{\beta_r(1/9, 10/9)}{\beta_l(1/9, 10/9)}$$

$$= \frac{33.61 \dot{q}_w \Pr^{0.4} R_x^{0.2}}{\varrho c_p u_e} \frac{\beta_r(1/9, 10/9)}{\beta_l(1/9, 10/9)},$$

which is Eq. (6.7.42) with $\beta_l(1/9, 10/9) = 8.844$.

6.15 PROBLEM

Show that the entrainment velocity v_E, which is the rate at which the volume flow rate per unit span changes with x, namely,

$$v_E = \frac{d}{dx} \int_0^\delta u \, dy$$

can be written in the form given by Eq. (6.9.1).

Hint: First use the continuity equation and Leibnitz's rule for differentiation to show that the v-component of the velocity at the boundary-layer edge, $y = \delta$, can be written as

$$v_e = -\frac{d}{dx} \int_0^\delta u \, dy + u_e \frac{d\delta}{dx}.$$

Then use the definition of displacement thickness δ^* to show that

$$v_e = \frac{d}{dx}(u_e \delta^*) - \delta \frac{du_e}{dx}$$

and

$$\int_0^\delta u \, dy = u_e(\delta - \delta^*).$$

SOLUTION

$$v_E \equiv \frac{d}{dx} \int_0^\delta u \, dy = \frac{d}{dx}\left[\int_0^\delta u_e \, dy - \int_0^\delta (u_e - u)\, dy\right] = \frac{d}{dx}[u_e(\delta - \delta^*)]$$

$$\therefore \quad \frac{v_E}{u_e} = \frac{1}{u_e}\frac{d}{dx}[u_e(\delta - \delta^*)].$$

6.16 PROBLEM

Show that the entrainment velocity v_E in axisymmetric flow is given by

$$v_E = \frac{1}{r_\delta}\frac{d}{dx}\int_0^\delta ru\,dy = \frac{1}{r_\delta}\frac{d}{dx}\left[u_e\left(\delta\frac{r_0+r_\delta}{2} - \delta^*\right)\right]$$

where

$$r_\delta = r_0 + \delta\cos\phi$$

$$\delta^* = \int_0^\delta r\left(1 - \frac{u}{u_e}\right)dy.$$

SOLUTION

$$v_E \equiv \frac{1}{r_\delta}\frac{d}{dx}\int_0^\delta ru\,dy = \frac{1}{r_\delta}\frac{d}{dx}\int_0^\delta r(u_e - u_e + u)\,dy$$

$$= \frac{1}{r_\delta}\frac{d}{dx}\left[\int_0^\delta u_e r\,dy - u_e\int_0^\delta r(1-u/u_e)\,dy\right]$$

where

$$r = r_0 + y\cos\phi(x), \qquad \delta^* \equiv \int_0^\delta r(1-u/u_e)\,dy$$

then

$$v_E = \frac{1}{r_\delta}\frac{d}{dx}\left[u_e\delta\left(r_0 + \frac{\delta}{2}\cos\phi(x)\right) - \delta^* u_e\right]$$

$$= \frac{1}{r_\delta}\frac{d}{dx}\left[u_e\delta\left(\frac{r_0}{2} + \frac{r_0 + \delta\cos\phi(x)}{2}\right) - \delta^* u_e\right];$$

since $r_\delta = r_0 + \delta\cos\phi(x)$

$$v_E = \frac{1}{r_\delta}\frac{d}{dx}\left[u_e\left(\frac{r_0+r_\delta}{2} - \delta^*\right)\right].$$

6.17 PROBLEM

The assumption of an adiabatic wall is appropriate to a glass windshield but less so to lifting surfaces of aircraft. Suppose that it is required to maintain a surface downstream of a blown slot at a uniform temperature of 300 °K by controlling the heat transfer to the surface with the assistance of electrical heaters. With the same parameters as those of Example 6.7, plot the distribution of heat flux over a 1 m length in dimensional terms. Note that g'_w, is defined by Eq. (4.7.12).

SOLUTION

$\nu = 14.2\times 10^{-6}$ and $k = 0.0252\,\text{W/m K}$ for $T_f = \frac{1}{2}(T_w+T_e) = 287.5\,\text{K}$ since $L = y_c = 0.0297\,\text{m}$, $\xi = (x-x_0)/y_c$,

$$R_L = \frac{u_e y_c}{\nu} = \frac{50.0 \times 0.0297}{14.2 \times 10^{-6}} = 1.046 \times 10^5, \quad T_c = 325\,\text{K}, \quad T_e = 275\,\text{K},$$

then

$$\dot{q}_w = \frac{g'_w(T_c - T_e)R_L^{1/2}k}{\xi^{1/2}L} = \frac{g'_w}{\xi^{1/2}} \times 50 \times (1.046 \times 10^5)^{1/2} \times \frac{0.0252}{0.0297}$$

$$= 1.37 \times 10^4 \frac{g'_w}{\xi^{1/2}}.$$

$x - x_0$	ξ	g'_w	$\dot{q}_w$
0	0	0	—
0.25	8.5	−0.01	−47.0
0.30	10.2	−0.02	−85.8
0.40	13.56	−0.11	−409.2
0.50	17.0	−0.24	−797.5
0.60	20.3	−0.38	−1155.5
0.70	23.7	−0.48	−1350.8
0.80	27.1	−0.60	−1579.0
0.90	30.5	−0.69	−1711.7
1.00	33.9	−0.78	−1835.3

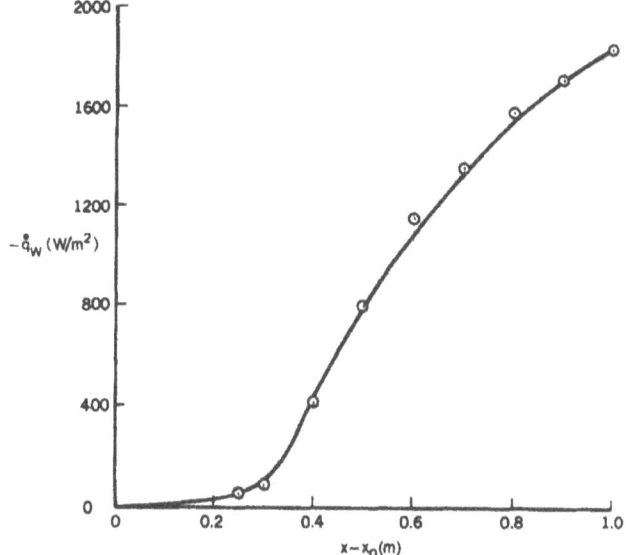

6.18 Problem

Air at atmospheric measure and $15\,^\circ\text{C}$ is heated as it flows over a two-dimensional smooth surface with an elliptic cross section of 4 to 1 thickness ratio. Calculate the wall shear parameter $c_f/2\sqrt{R_x}$ and wall heat flux parameter $\mathrm{Nu}_x/\sqrt{R_x}$ as a function of dimensionless axial distance $x/2a$ for both laminar and turbulent flows with transition computed by Michel's method for $R_{2a} = 10^7$ and for a uniform wall temperature of $25\,°\mathrm{C}$ and for a Prandtl number 0.72. Compare your results with those given in Figures 6.15 and 6.16.

SOLUTION

Consider a flow of air which is initially laminar and becomes turbulent around a two-dimensional ellipse of thickness ratio 4, with $T_\infty = 15\,°\mathrm{C}$ and $T_w = 25\,°\mathrm{C}$, $R_{2a} = 10^7$, $\mathrm{Pr} = 0.72$. The location of transition is obtained by Michel's method where $R_{x_{tr}}$ is the Reynolds number based on the surface distance. Laminar and turbulent flow calculations are performed with a procedure similar to one used for Test Case 3 in subsection 11.8.5. The results for $c_f/2\sqrt{R_x}$ vs x/a and $\mathrm{Nu}_x/\sqrt{R_x}$ vs (x/a) are shown below.

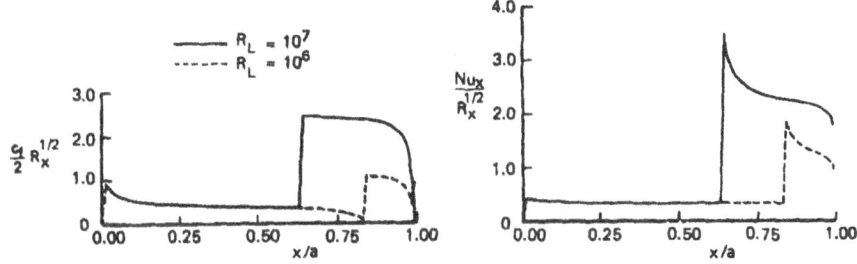

7
Turbulent Duct Flows

7.1 PROBLEM
Deduce Eq. (7.2.6) from Eq. (7.2.5).

SOLUTION

Since
$$\tau_w = \varrho u_\tau^2 = \mu \left(\frac{du}{dy}\right)_{y=0} = -\mu \left(\frac{du}{dr}\right)_{r=r_0} \quad \text{or} \quad \left(\frac{du}{dr}\right)_{r=r_0} = -\frac{u_\tau^2}{\nu} \quad (1)$$

Eq. (7.2.5) can be written at $r = r_0$ as
$$r_0^K \left(\frac{du}{dr}\right)_{r=r_0} = \frac{r_0^{2K}}{K+1} \frac{1}{\nu} \frac{d\overline{P}}{dx}. \quad (2)$$

From Eqs. (1) and (2), it follows that
$$-r_0^K \frac{u_\tau^2}{\nu} = \frac{r_0^{2K}}{K+1} \frac{1}{\nu} \frac{d\overline{P}}{dx} = \left(\frac{r_0}{r}\right)^{2K} \frac{r^{2K}}{K+1} \frac{1}{\nu} \frac{d\overline{P}}{dx} = \left(\frac{r_0}{r}\right)^{2K} b \frac{du}{dr}$$
$$= -\left(\frac{r_0}{r}\right)^{2K} b \frac{du}{dy}. \quad (3)$$

With $b = r^K(1+\varepsilon_m^+)$, Eq. (3) can be written in the form of Eq. (7.2.6)
$$\left(\frac{r_0}{r}\right)^{2K} r^K (1+\varepsilon_m^+) \frac{du}{dy} = r_0^K \frac{u_\tau^2}{\nu} \quad \text{or} \quad (1+\varepsilon_m^+) \frac{du}{dy} = \left(\frac{r}{r_0}\right)^K \frac{u_\tau^2}{\nu}.$$

7.2 PROBLEM
Use the velocity defect law for a pipe, that is,
$$\frac{u_{\max} - u}{u_\tau} = \phi\left(\frac{y}{r_0}\right)$$

and the empirically obtained relation that

$$u_m = u_{\max} - 4.07 u_\tau$$

and show that the friction factor f can be written as

$$u_{\max} = u_m(1 + 1.44\sqrt{f}). \tag{P7.1}$$

SOLUTION

By definition

$$f = \frac{\tau_w}{(1/8)\varrho u_m^2} = \frac{u_\tau^2}{(1/8) u_m^2} \quad \text{or} \quad u_\tau = u_m\sqrt{f/8}.$$

If $u_m = u_{\max} - 4.07 u_\tau$, then

$$u_{\max} = u_m + 4.07 u_\tau = u_m(1 + 4.07\sqrt{f/8}) = u_m(1 + 1.44\sqrt{f}).$$

7.3 PROBLEM

Evaluate Eq. (7.2.14) at the centerline of the pipe and, with Eq. (P7.1), show that Prandtl's friction law for smooth pipes can be written in the form given by Eq. (7.2.15)

SOLUTION

In pipe flow Π is negligibly small; the velocity at $y = r_0$ can then be expressed as

$$u_{\max}^2 = \frac{1}{\kappa} \ln \frac{r_0 u_\tau}{\nu} + c = \frac{1}{\kappa} \ln \left(\frac{r_0 u_m}{\nu} \sqrt{f/8} \right) + c \tag{1}$$

where $u_{\max}^+$ is related to u_m^* by

$$u_{\max}^+ = u_m^+(1 + 1.44\sqrt{f}). \tag{2}$$

Combining (1) and (2),

$$u_m^+(1 + 1.44\sqrt{f}) = \frac{1}{\kappa} \ln \left(\frac{r_0 u_m}{\nu} \sqrt{f/8} \right) + c$$

or

$$\frac{1}{\sqrt{f/8}}(1 + 1.44\sqrt{f}) = \frac{1}{\kappa}(\ln \sqrt{f} R_d - 0.5 \ln 8) + c$$

or

$$\frac{1}{\sqrt{f}} = \frac{1}{\sqrt{8}}\frac{1}{\kappa}\ln(\sqrt{f} R_d) + \left[\frac{1}{\sqrt{8}}\left(-\frac{0.5}{\kappa}\ln 8 + c\right) - 1.44 \right].$$

Let

7. Turbulent Duct Flows

$$A = \frac{1}{\sqrt{8}}\frac{1}{\kappa}, \quad B = \frac{1}{\sqrt{8}}(c - \frac{0.5}{\kappa}\ln 8) - 1.44$$

then

$$\frac{1}{\sqrt{f}} = A \ln R_d \sqrt{f} + B$$

which is in the form of Prandtl's friction law for smooth pipes. For $\kappa = 0.41$ and $c = 5.2$, $A = 0.86$ and $B = -0.5$. For better agreement with the experimental data, set $A = 0.87$ and $B = -0.8$, which is the Prandtl law given by (7.2.15).

7.4 PROBLEM

Show that Prandtl's friction law for a two-dimensional duct, corresponding to the pipe law in Problem 7.3, can be written, with $(R = u_m h/\nu)$, as

$$\frac{1}{\sqrt{f}} = 0.87 \ln(R\sqrt{f}) - 0.41.$$

SOLUTION

Again, since Π is negligibly small in channel flow, the entire velocity profile can be represented by $u^+ = (1/\kappa)\ln(yu_\tau/\nu) + c$ and be used to find $u_{\max}$ and u_m.

$$u_{\max}^+ = \frac{1}{\kappa}\ln\frac{hu_\tau}{\nu} + c, \quad u_m^+ = \frac{1}{\kappa}\left(\ln\frac{hu_\tau}{\nu} - 1.0\right) + c \tag{1}$$

$$\therefore \ u_m^+ = u_{\max}^+ - \frac{1}{\kappa}. \tag{2}$$

From the second equation in (1), we can write,

$$\frac{1}{\sqrt{f/8}} = \frac{1}{\kappa}(\ln\sqrt{f/8}R - 1.0) + c$$

$$\frac{1}{\sqrt{f}} = \frac{1}{\sqrt{8}}\frac{1}{\kappa}(\ln\sqrt{f}R - 0.5\ln 8 - 1.0) + \frac{c}{\sqrt{8}}$$

$$= \frac{1}{\sqrt{8}}\frac{1}{\kappa}(\ln\sqrt{f}R) + \frac{c}{\sqrt{8}} - \frac{1}{\sqrt{8}\kappa}(0.5\ln 8 + 1.0).$$

Let

$$\frac{1}{\kappa}\frac{1}{\sqrt{8}} = A \quad \text{and} \quad \frac{1}{\sqrt{8}}[c - (0.5\ln 8 + 1.0)/\kappa] = B.$$

Then

$$\frac{1}{\sqrt{f}} = A \ln R\sqrt{f} + B$$

is in the form of Prandtl's friction law for turbulent flow in a smooth channel. With $\kappa = 0.41$ and $c = 5.2$, $A = 0.86$ and $B = 0.08$. For better agreement with experimental data, set $A = 0.87$, $B = -0.41$, we have Prandtl's friction law, expressed as

$$\frac{1}{\sqrt{f}} = 0.87 \ln R\sqrt{f} - 0.41.$$

7.5 PROBLEM

Integrate Eq. (7.2.18) with respect to r twice subject to the boundary conditions given by Eq. (7.2.20) for the case of turbulent flow with constant heat flux condition at the wall. Use the mixing-length formula given by Eqs. (7.2.8) and (7.2.9) and the turbulent Prandtl number expression given by Eq. (6.4.19) for $R_d = 10^4$ and 10^5 with $\text{Pr} = 0.02, 0.72$ and 14.3. Compute the friction factor by using Eq. (7.2.16). Compare your results with those given in Fig. 7.2.

SOLUTION

See the computer program in the accompanying CD for $R_d = 10^4$, $\text{Pr} = 0.02$. Computed results for uniform heat flux at two Reynolds numbers are given below.

Pr	Nu ($R_d = 10^4$)	Nu ($R_d = 10^5$)
0.02	6.78	15.61
0.72	29.00	172.69
14.3	106.07	831.66

7.6 PROBLEM

Repeat Problem 7.5 for uniform wall temperature.

SOLUTION

See the computer program in the accompanying CD for $R_d = 10^5$, $\text{Pr} = 14.3$. Computed results for the uniform wall temperature at two Reynolds numbers are given below.

Pr	Nu ($R_d = 10^4$)	Nu ($R_d = 10^5$)
0.02	5.39	13.80
0.72	27.89	169.39
14.3	105.75	830.59

7.7 PROBLEM

Compare the values of Nusselt number obtained from Eqs. (7.2.22), (7.2.23) and (7.2.25) for the following pipe flows with

(a) $Pr = 0.7$, $R_d = 10^4$ (Compare Eq. (7.2.24) also),
(b) $Pr = 0.2$, $R_d = 10^4$,
(c) $Pr = 100$, $R_d = 10^4$,
(d) $Pr = 0.7$, $R_d = 10^6$.

SOLUTION

The use of Eqs. (7.2.22), (7.2.23), (7.2.24) and (7.2.25) for evaluating the Nusselt number yield:

a. $Pr = 0.7$, $R_d = 10{,}000$, $f = 0.3164/R_d^{0.25} = 0.03164$

	Eq. (7.2.22)	Eq. (7.2.23)	Eq. (7.2.24)	Eq. (7.2.25)
Nu	39.95	36.4	32.7	53.3

b. $Pr = 0.2$, $R_d = 10.000$

	Eq. (7.2.22)	Eq. (7.2.23)	Eq. (7.2.25)
Nu	14.6	15.7	24.4

c. $Pr = 100$, $R_d = 10.000$

	Eq. (7.2.22)	Eq. (7.2.23)	Eq. (7.2.25)
Nu	140.6	3045.0	221.9

d. $Pr = 0.7$, $R_d = 10^6$, $f = 0.3164/R_d^{0.25} = 0.01$

	Eq. (7.2.22)	Eq. (7.2.23)	Eq. (7.2.25)
Nu	390.0	1515.2	401.4

7.8 Problem

Liquid mercury enters a 0.02 m-diameter tube at 50 °C and is heated to 95 °C as it passes through the tube at a mass flow rate of $2\,\text{kg s}^{-1}$. If a constant heat flux is maintained along the tube and the surface temperature of the tube is 15 °C higher than the liquid mercury bulk temperature, calculate the length of the tube.

SOLUTION

$T_m = 0.5(T_e + T_w) = 0.5(50 + 110.0) = 80\,°C$. For liquid mercury with

$$T_m = 80\,°C, \quad \varrho = 13434.6\,\text{kg/m}^3, \quad \nu = 0.097 \times 10^{-6}\,\text{m}^2/\text{s},$$
$$\kappa = 5.42 \times 10^{-6}\,\text{m}^2/\text{s}, \quad Pr = 0.018.$$

Mass flux

$$\dot{m} = \varrho u_m A = 13434.58 \times \pi \times (0.01)^2 u_m = 2.0.$$

$$\therefore \quad u_m = 0.474\,\text{m s}^{-1} \quad \text{and} \quad R_d = \frac{u_m d}{\nu} = \frac{0.474 \times 0.02}{0.097 \times 10^{-6}} = 0.98 \times 10^5.$$

For a low Prandtl number fluid, we calculate Nu from the formula proposed by Sleicher and Rouse, which is:

$$\mathrm{Nu} = 6.3 + 0.0167 R_d^{0.85} \mathrm{Pr}^{0.93} = 6.3 + 0.0167(0.98 \times 10^5)^{0.85}(0.018)^{0.93}$$
$$= 13.24.$$

The total heat flux is

$$Q = \int_0^L \dot{q}_w p\, dx = \pi d L \dot{q}_w = \pi d L (T_w - T_m)\mathrm{Nu}\frac{k}{d} = (T_w - T_m)\mathrm{Nu}\pi k L$$
$$= c_p(T_0 - T_i)\varrho u_m \frac{\pi}{4} d^2$$

$$\therefore\ L = \frac{\varrho c_p}{k} \frac{T_0 - T_i}{T_w - T_m} \frac{u_m d^2 \mathrm{Nu}^{-1}}{4}$$
$$= \left(54.2 \times 10^{-7}\right)^{-1} \frac{45}{15} \frac{0.474}{4} \times 0.02^2 \times 13.24^{-1} = 1.98\,\mathrm{m/s}.$$

7.9 PROBLEM

Water at a rate of $1\,\mathrm{kg\,s^{-1}}$ and at an average temperature of $20\,^\circ\mathrm{C}$ flows in a $0.05\,\mathrm{m}$-diameter tube $15\,\mathrm{m}$ long. The pressure drop is measured as $5 \times 10^{-6}\,\mathrm{Pa\,s}$. A uniform heat flux condition is maintained at the wall and the average wall temperature is $50\,^\circ\mathrm{C}$ and is $10\,^\circ\mathrm{C}$ higher than the water bulk temperature. Calculate the exit temperature of the water. Use the computer program of Problem 11.4.

SOLUTION

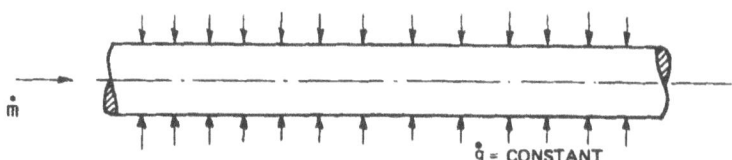

$\dot{m} = 1\,\mathrm{kg/s}$, $T = 20\,^\circ\mathrm{C}$, $d = 5\times 10^{-2}\,\mathrm{m}$, $P = 15\,\mathrm{m}$, $T_w = 50\,^\circ\mathrm{C}$, $T_w - T_m = 10\,^\circ\mathrm{C}$. Evaluate properties of water at $20\,^\circ\mathrm{C} = 273 + 20 = 293\,\mathrm{K}$.

$$\varrho = 1000\,\mathrm{kg/m^3}, \quad c_p = 4.1818\,\mathrm{kJ/kg\,K}, \quad \nu = 1.006 \times 10^{-6}\,\mathrm{m^2/s},$$
$$k = 0.597\,\mathrm{W/m\,K}, \quad \mathrm{Pr} = 7.02.$$

To find the Reynolds number R_{r_0} and R_d

$$\bar{u} = \frac{4\dot{m}}{\varrho \pi d^2} = \frac{4 \times 1}{10^3 \times \pi \times 25 \times 10^{-4}}, \quad \bar{u} = \frac{40}{25 \times \pi} = 0.5\,\mathrm{m/s}$$
$$R_{r_0} = \frac{2.5 \times 10^{-2} \times 0.5}{1.006 \times 10^{-6}} = 12425, \quad R_d = 25{,}000$$

l	$\hat{x}$	x/d	Nu_i	$h_i = (\mathrm{Nu}_i k)/d$
$l = 15\,\mathrm{m}$	$\hat{x} = 6.84 \times 10^{-3}$	$x/d = 300$	174.5	$2083.5\,\mathrm{W/m^2\,K}$
$l = 10\,\mathrm{m}$	$\hat{x} = 4.56 \times 10^{-3}$	$x/d = 200$	174.5	$2083.5\,\mathrm{W/m^2\,K}$
$l = 5\,\mathrm{m}$	$\hat{x} = 2.28 \times 10^{-3}$	$x/d = 100$	174.5	$2083.5\,\mathrm{W/m^2\,K}$
$l = 2.5\,\mathrm{m}$	$\hat{x} = 1.14 \times 10^{-3}$	$x/d = 50$	174.5	$2083.5\,\mathrm{W/m^2\,K}$
$l = 1.0\,\mathrm{m}$	$\hat{x} = 4.56 \times 10^{-4}$	$x/d = 20$	175.05	$2090\,\mathrm{W/m^2\,K}$
$l = 0.5\,\mathrm{m}$	$\hat{x} = 2.28 \times 10^{-4}$	$x/d = 10$	177.6	$2120.5\,\mathrm{W/m^2\,K}$
$l = 0.25\,\mathrm{m}$	$\hat{x} = 1.14 \times 10^{-4}$	$x/d = 5$	181.9	$2171.88\,\mathrm{W/m^2\,K}$
$l = 0.1\,\mathrm{m}$	$\hat{x} = 4.55 \times 10^{-5}$	$x/d = 2$	191.2	$2283.0\,\mathrm{W/m^2\,K}$

We use the computer program in Problem 11.4 to obtain the Nusselt number distributions and find that $\mathrm{Nu}_\infty = 174.0$ for $x/d \geq 20$.

$$\dot{m} c_p (T_\mathrm{exit} - T_\mathrm{inlet}) = \dot{q}_\mathrm{total} = h(\pi d L)(T_w - T_\mathrm{bulk})$$

$$\therefore\ T_\mathrm{exit} = T_\mathrm{inlet} + \frac{h(\pi d l)}{\dot{m} c_p}(T_w - T_\mathrm{bulk})$$

$$= 20\,^\circ\mathrm{C} + \frac{\pi d}{\dot{m} c_p}(T_w - T_\mathrm{bulk}) \sum_{i=1}^{N} h_i l_i$$

$$= 20 + 3.756 \times 10^{-4} \sum_{i=1}^{N} h_i l_i.$$

Here h_i and l_i denote the heat tranfer coefficient and the pipe length over the ith segment, respectively

$$T_\mathrm{exit} = 20 + 3.756 \times 10^{-4}(2283 \times 0.1 + 2171.88 \times 0.15$$
$$+ 2120.5 \times 0.25 + 2090 \times 0.5 + 2083.5 \times 14)$$
$$T_\mathrm{exit} = 20 + 3.756 \times 10^{-4}(228.3 + 325.8 + 530.1 + 1045 + 29169)$$
$$= 20\,^\circ\mathrm{C} + 11.75\,^\circ\mathrm{C} = 31.75\,^\circ\mathrm{C}.$$

7.10 Problem

Repeat Problem 7.5 for a turbulent flow in a pipe with roughness, $r_0/k_s = 100$ for $R_d = 10^5$, $\mathrm{Pr} = 0.02$, 0.72 and 14.3. Compute the friction factor from Eq. (7.2.27) and compare your results with those given in Fig. 7.8.

Hint: To calculate the velocity profile, use the two-layer eddy-viscosity model in which the inner-region eddy viscosity is obtained from Eq. (7.2.9) with the mixing length l given by Eq. (10.3.1), and the outer-region eddy viscosity from Eq. (7.2.11) with the value of parameter α obtained either in Fig. 7.1 or by making use of Eq. (7.2.17). Since the establishment of inner and outer regions requires a velocity profile, obtain the initial profile from Eq. (7.2.12) with the modified mixing-length given by Eqs. (7.2.8) and (7.2.13). Then compute the eddy-viscosity distribution from the two-layer model and the velocity profile from Eq. (7.2.7). Repeat this procedure until the computed velocity profiles converge.

SOLUTION

Modify the computer program of Problem 7.5 for the case of rough pipe to include the two-layer eddy-viscosity model defined separately in the inner and outer layers of the shear layer, and as given by Eqs. (7.2.9), (10.3.8), (10.3.9) and (7.2.11) with $\alpha = 0.0168$. The switch from $(\varepsilon_m)_i$ to $(\varepsilon_m)_o$ is achieved be the continuity in ε_m, namely $\varepsilon_m = (\varepsilon_m)_i$ if $(\varepsilon_m)_i \leq (\varepsilon_m)_o$ and $\varepsilon_m = (\varepsilon_m)_o$ when $(\varepsilon_m)_o > (\varepsilon_m)_i$. As in Problem 7.5 the initial velocity profile is obtained by making use of the modified mixing length model. However, this time the friction factor, f, was calculated from Eq. (7.2.27) in order to account for the effect of the surface roughness parameter (r_0/k_s). The velocity distribution u^+ is obtained by integrating Eq. (7.2.12), and the velocity profile is computed until the $u(y)$ distribution satisfies the continuity equation given by Eq. (7.2.17).

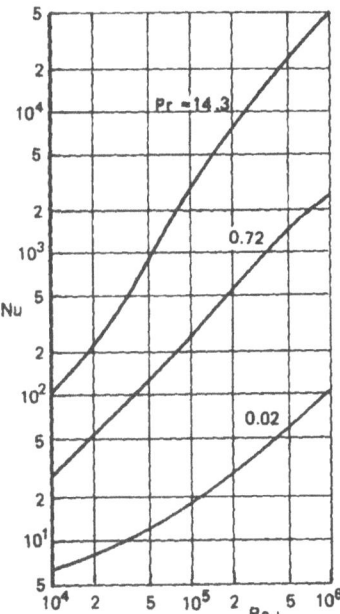

For the case of rough surface, the mixing-length expression is defined according to Eq. (10.3.8) with Δy expressed as a function of an equivalent sand grain-roughness parameter, k_s^+ given by Eq. (10.3.9).

The values of the Nusselt number obtained with the modified computer program given below for $R_d = 10^5$ are:

7. Turbulent Duct Flows

Pr	$(Nu)_{Fig. 7.8^*}$	$(Nu)_{computed}$
14.3	3300	2981
0.72	300	250
0.02	20.0	18.02

* Actually from Problem 11.3.

7.11 PROBLEM

Water at a rate of $3\,\mathrm{kg\,s^{-1}}$ and at $40\,^\circ\mathrm{C}$ enters a 5 cm-diameter pipe with a relative roughness of $k_s/d = 0.005$. If the wall temperature is maintained at $60\,^\circ\mathrm{C}$ and is $10\,^\circ\mathrm{C}$ higher than the bulk water temperature, and the pipe is 20 m long, calculate the total heat transfer. Use Reynolds analogy.

SOLUTION

$T_f = 0.5(T_e + T_w) = 0.5(40 + 60) = 50\,^\circ\mathrm{C}$. The fluid properties of water at $T_f = 50\,^\circ\mathrm{C}$ are: $\varrho = 990\,\mathrm{kg/m^3}$, $c_p = 4.18\,\mathrm{kJ/kg\,K}$, $\nu = 0.568 \times 10^{-6}\,\mathrm{m^2/s}$, $k = 0.640\,\mathrm{W/m\,K}$, $\kappa = 0.153 \times 10^{-6}\,\mathrm{m^2/s}$, $\mathrm{Pr} = 3.71$. Mass flow rate, $\dot{m} = \varrho u_m A = 990 u_m \pi (0.05^2/4) = 3$.

$$\therefore \quad u_m = 1.543\,\mathrm{m/s} \quad \text{and} \quad R_d = \frac{u_m d}{\nu} = \frac{0.05 \times 1.543}{0.568 \times 10^{-6}} = 1.36 \times 10^5.$$

$$f = \frac{1}{[(2\log_{10}(r_0/k_s) + 1.74]^2} = \frac{1}{[2\log_{10}(100) + 1.74]^2} = 0.03035.$$

To compute St, we use Reynolds analogy, $\mathrm{St}/(f/8) = 1.0$.

$$\therefore \quad \mathrm{St} = f/8 = 0.00379$$

so that

$$\mathrm{Nu} = \mathrm{Pr}\,R_d\,\mathrm{St} = 3.71 \times 1.36 \times 10^5 \times 0.003790 = 1914$$

and

$$\dot{q}_w = \mathrm{Nu}(T_w - T_m)\frac{k}{d}\pi dL = 1914 \times 10 \times 0.640 \times \pi \times 20 = 7.7 \times 10^5\,\mathrm{W}.$$

7.12 PROBLEM

$0.5\,\mathrm{kg\,s^{-1}}$ of air at atmospheric pressure and $10\,^\circ\mathrm{C}$ is heated as it flows through a 2 cm-diameter pipe with a relative roughness of $k_s/d = 0.002$. Calculate the heat transfer per unit length of pipe if a constant-heat-flux condition is maintained at the wall and the wall temperature is $5\,^\circ\mathrm{C}$ above the average air temperature.

SOLUTION

The fluid properties of air at $T = 10\,°C$ are $\varrho = 1.257\,\mathrm{kg/m^3}$, $c_p = 1.0055\,\mathrm{kJ/kg\,K}$, $\nu = 1.36 \times 10^{-5}\,\mathrm{m^2/s}$, $k = 0.0249\,\mathrm{W/m\,K}$, $\mathrm{Pr} = 0.72$. The friction factor is calculated from the formula

$$f = \frac{1}{[2\log_{10}(r_0/k_s) + 1.74]^2} = 0.0234$$

and R_d is

$$R_d = \frac{ud}{\nu} = \frac{\dot{m}d}{\varrho A v} = \frac{0.5 \times 0.02}{1.257 \times \pi \times (0.01)^2 \times 1.36 \times 10^{-5}} = 1.87 \times 10^6.$$

Assume that the Reynolds analogy is valid, i.e., $\mathrm{St}/(f/8) = 1.0$, so that $\mathrm{Nu} = \mathrm{Pr} R_d \mathrm{St} = 0.72 \times 1.87 \times 10^6 \times (0.0234/8) = 3.94 \times 10^3$, and $\dot{q}_w = \mathrm{Nu}(T_w - T_m)k/d = 3.94 \times 10^3 \times 5 \times 0.0249/0.02 = 2.45 \times 10^4\,\mathrm{W/m^2}$. The heat transfer per unit length of the pipe is then

$$\dot{Q}_w = \dot{q}_w \cdot 2\pi r = 2.45 \times 10^4 \times 2\pi \times 0.01 = 1.54\,\mathrm{W/m}.$$

7.13 PROBLEM

7.13 Air at atmospheric pressure and $10\,°C$ flows in a 0.05 m-diameter, 1 m long pipe at a velocity of $10\,\mathrm{m\,s^{-1}}$. The heating starts at 5 cm from the entrance. A constant heat flux is imposed such that the wall temperature is $5\,°C$ above the average air temperature. Calculate the local heat transfer rates at 13, 25 and 45 cm from the entrance of the tube. Assume velocity field is fully developed.

SOLUTION

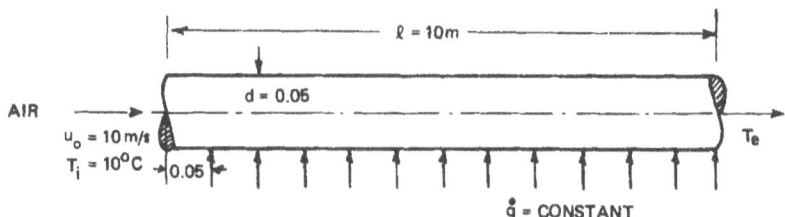

Assume fully developed velocity field and evaluate the properties of air at $10\,°C$. $\nu = 12.58 \times 10^{-6}\,\mathrm{m^2/s}$; $k = 0.02624\,\mathrm{W/m\,K}$ then

$$R_{r_0} = \frac{u_0 r_0}{\nu} = \frac{(10\,\mathrm{m/s})(2.5 \times 10^{-2})\,\mathrm{m}}{12.58 \times 10^{-6}\,\mathrm{m^2/s}} = 20 \times 10^3.$$

With constant heat flux boundary condition, Eqs. (P7.2) and (P7.3), with $\mathrm{Nu}_\infty = 88.47$ and $R_d = 40 \times 10^3$:

$$\frac{\mathrm{Nu}}{\mathrm{Nu}_\infty} = 1.0 + 0.8(1.0 + 70{,}000 R_d^{-1.5})\left(\frac{x}{d}\right)^{-1},$$

$$\mathrm{Nu}_\infty = 0.021 \times R_d^{0.8} \times \mathrm{Pr}^{0.4}$$

x/d	$\mathrm{Nu}/\mathrm{Nu}_\infty$	Nu	$\bar{h} = \mathrm{Nu}\,k/d$	$\dot{q}_w = \bar{h}(T_w - T)$
$8/5 = 1.6$	1.50	132.7	69.64 W/m² °C	348.2 W/m²
$20/5 = 4.0$	1.20	106.2	55.7 W/m² °C	278.5 W/m²
$40/5 = 8.0$	1.10	97.3	51.1 W/m² °C	255.4 W/m²

7.14 PROBLEM

$2\,\mathrm{kg\,s^{-1}}$ of air at atmospheric pressure flows in a 0.1 m-diameter, 1.5 m long pipe. A constant uniform wall temperature of 140 °C is applied for 1 m starting 0.5 m from the entrance. The average air temperature in the pipe is 200 °C. Calculate

(a) the local heat transfer rates at 0.3, 0.4 and 0.6 m.
(b) the decrease in temperature of the air as it passes through the duct.

SOLUTION

Evaluate the properties of air at $\overline{T} = 200\,°\mathrm{C}$ (473 K). Pr = 0.680; $\nu = 53.38 \times 10^{-6}\,\mathrm{m^2/s}$; $k = 0.0387\,\mathrm{W/m\,K}$; $\varrho = 0.744\,\mathrm{kg/m^3}$; $c_p = 1.024\,\mathrm{kJ/kg\,K}$.

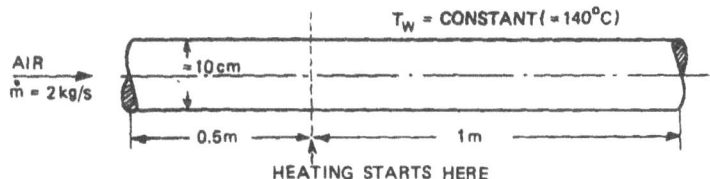

Since $\dot{m} = \varrho \bar{u}\frac{\pi d^2}{4}$, $\bar{u} = \frac{4\dot{m}}{\varrho \pi d^2} = \frac{4 \times 2}{0.744 \times \pi \times 0.1^2} = 3.42 \times 10^2\,\mathrm{m\,s^{-1}}$.
$R_{r_0} = \frac{\bar{u}r_0}{\nu} = \frac{342\,\mathrm{m/s}(0.05)}{53.38 \times 10^{-6}} = 512{,}282$; Take $R_{r_0} = 500{,}000$ or $R_d = 10^6$.
At this Reynolds number, from Fig. 7.2, $\mathrm{Nu}_\infty \to 1150$. Also use Fig. 7.9b to obtain the variation of $\mathrm{Nu}/\mathrm{Nu}_\infty$ with x/d; after the heating starts. From Fig. 7.9.b; at $R_d = 10^6$.

$$\frac{\mathrm{Nu}}{\mathrm{Nu}_\infty}\left(\frac{x}{d} = \frac{0.3}{0.1}\right) \to \frac{\mathrm{Nu}}{\mathrm{Nu}_\infty}\left(\frac{x}{d} = 3\right) = 1.2,$$

$$\mathrm{Nu} = 1380;\quad h = 534.1\,\mathrm{W/m^2\,K}.$$

$$\frac{\mathrm{Nu}}{\mathrm{Nu}_\infty}\left(\frac{x}{d} = 4\right) \to \frac{\mathrm{Nu}}{\mathrm{Nu}_\infty}\left(\frac{x}{d} = 4\right) = 1.15,$$

$$\mathrm{Nu} = 1322.5;\quad h = 511.8\,\mathrm{W/m^2\,K}.$$

$$\frac{\mathrm{Nu}}{\mathrm{Nu}_\infty}\left(\frac{x}{d}=6\right) \to \frac{\mathrm{Nu}}{\mathrm{Nu}_\infty}\left(\frac{x}{d}=6\right) = 1.10,$$

$$\mathrm{Nu} = 1265; \quad h = 490\,\mathrm{W/m^2\,K}.$$

The local heat transfer rates are then:

x/d	$h\,[\mathrm{W/m^2\,K}]$	$\dot{q}_w = h(T_w - T)\,[\mathrm{W/m^2}]$	ΔT (Temperature drop)
3	534	-32040	$-1.47\,^\circ\mathrm{C}$
4	512	-30720	$-0.47\,^\circ\mathrm{C}$
6	490	-29400	$-0.90\,^\circ\mathrm{C}$

The decrease in air temperature is obtained from

$$\dot{m}c_p(T_i - T(x)) = h\pi d(x)(T_w - T), \quad T_i - T(x) = \frac{h\pi dx(T_w - T)}{\dot{m}c_p}$$

or

$$\Delta T_1 = T(x=0.3) - T_i = \frac{(-32040)(\pi)(0.1)(0.3)}{(2)(1024)} = -1.47\,^\circ\mathrm{C}.$$

Similarly $\Delta T_2 = -0.47\,^\circ\mathrm{C}$, $\Delta T_3 = -0.90\,^\circ\mathrm{C}$.
$\Delta T_{\mathrm{total}} = -2.84\,^\circ\mathrm{C}$, total temperature drop $\approx 3\,^\circ\mathrm{C}$.

7.15 Problem

As a crude approximation, the thermal entry length of a turbulent flow in a circular duct with uniform surface temperature and $R_d = 10^4$, $\mathrm{Pr} = 100$ can be estimated from the Leveque analysis discussed in Problem 4.15. To do this, assume that the Leveque solution is valid up to the fully-developed limit, for which the Nusselt number may be obtained from appropriate fully-developed analysis. Put your results in terms of number of diameters from the start of heating.

Solution

Use the Leveque solution given in Problem 4.15 where

$$\mathrm{Nu}_x = \frac{x}{0.893}\left(\frac{\lambda \mathrm{Pr}}{9\nu x}\right)^{1/3} \quad \text{with} \quad \lambda = \left(\frac{\partial u}{\partial y}\right)_w.$$

Since near wall, $u^+ = y^+$ and $du^+/dy^+ = 1$

$$\left(\frac{\partial u}{\partial y}\right)_w = \frac{u_\tau^2}{\nu} = \frac{1}{\nu}\left(\frac{\tau_w}{\varrho}\right).$$

From the definition of friction factor f and Eq. (7.2.16), it follows that

7. Turbulent Duct Flows

$$\left(\frac{\partial u}{\partial y}\right)_w = \frac{1}{\nu}\frac{f}{8}u_m^2 = \frac{1}{\nu}\frac{0.3164}{8R_d^{1/4}}u_m^2.$$

Substituting the above expression into the Nu_x-expression, with $\overline{x} = \frac{x}{d}$,

$$\mathrm{Nu}_x = \frac{\overline{x}d}{0.893}\left(\frac{1}{\nu^2}\frac{\mathrm{Pr}}{9\overline{x}d}\frac{0.3164}{R_d^{1/4}}u_m^2\right)^{1/3}$$

and rearranging

$$\mathrm{Nu}_x = \frac{1}{0.893}\left(0.004\frac{R_d^2\mathrm{Pr}\overline{x}^1}{R_d^{1/4}}\right)^{1/3}$$

Substituting the numerical values

$$\mathrm{Nu}_x = \frac{1}{0.893}\left(0.0044\frac{(10^4)^2 \times 100\overline{x}^1}{(10^4)^{1/4}}\right)^{1/3} = 187 \times \overline{x}^{1/3}$$

Since, with l_t denoting the entrance length,

$$\overline{\mathrm{Nu}} = \frac{1}{l_t}\int_0^{l_t}\mathrm{Nu}(x)\,dx = \frac{d}{l_t}(187)\frac{(x^+)^{2/3}}{2/3}\bigg|_0^{l_t/d} = 281\left(\frac{l_t}{d}\right)^{1/3}.$$

From Fig. 7.5, $\overline{\mathrm{Nu}} = 200$

$$\therefore \quad l_t/d = 22.$$

8. Buoyant Flows

8.1 PROBLEM

Show that the square of the ratio

$$\frac{u_c}{u_e} = \frac{\sqrt{g\beta L(T_w - T_e)}}{u_e}$$

is a form of the Richardson number defined in Eq. (8.1.1).

SOLUTION

From the definition $\beta = -\frac{1}{\varrho}\left(\frac{\partial \varrho}{\partial T}\right)_p$, we can write

$$\Delta T \beta = -\frac{\Delta \varrho}{\varrho} \quad \text{or} \quad (T_w - T_e)\beta = -\frac{(\varrho_e - \varrho_w)}{\varrho}.$$

Substituting the above relation into

$$\frac{u_c}{u_e} = \frac{\sqrt{g\beta L(T_w - T_e)}}{u_e},$$

we get

$$\left(\frac{u_c}{u_e}\right)^2 = \frac{\Delta \varrho}{\varrho}\frac{gL}{u_e^2},$$

which, by definition, is the Richardson number.

8.2 PROBLEM

Derive the similarity equations for two-dimensional laminar natural convection flow over a vertical flat plate subject to (a) specified surface temperature and (b) heat flux. Use the transformation defined by Eq. (8.2.2).

Solution

For 2-D laminar natural convection flow over a vertical flat plate, the governing equations and their boundary conditions are

$$\frac{\partial u}{\partial x} + \frac{\partial v}{\partial y} = 0$$

$$u\frac{\partial u}{\partial x} + v\frac{\partial u}{\partial y} = \nu\frac{\partial^2 u}{\partial y^2} + g\beta(T - T_e)$$

$$u\frac{\partial T}{\partial x} + v\frac{\partial T}{\partial y} = \frac{\nu}{\Pr}\frac{\partial^2 T}{\partial y^2}$$

$y = 0, \quad u = v = 0, \quad T = T_w \quad \text{or} \quad \frac{\partial T}{\partial y} = -\frac{\dot{q}_w}{k}; \quad y = \delta, \quad u = 0, \quad T = T_e$

a. For specified surface temperature, use the transformation

$$\eta = \left[\frac{g\beta(T_w - T_e)}{\nu^2 x}\right]^{1/4} y, \quad \psi = [g\beta(T_w - T_e)\nu^2 x^3]^{1/4} f(\eta). \quad (5)$$

Then

$$u = \left(\frac{\partial \psi}{\partial y}\right)_x = \frac{\partial \psi}{\partial \eta}\frac{\partial \eta}{\partial y} = [g\beta(T_w - T_e)x]^{1/2} f'(\eta)$$

$$v = -\left(\frac{\partial \psi}{\partial x}\right)_y = -\left(\frac{\partial \psi}{\partial x}\right)_\eta - \frac{\partial \psi}{\partial \eta}\frac{\partial \eta}{\partial x}$$

$$= -[g\beta(T_w - T_e)\nu^2 x^3]^{1/4}\left[\frac{1}{4}(n+3)\frac{f}{x} + f'\frac{\partial \eta}{\partial x}\right]$$

$$\left(\frac{\partial u}{\partial x}\right)_y = \left(\frac{\partial u}{\partial x}\right)_\eta + \frac{\partial u}{\partial \eta}\frac{\partial \eta}{\partial x}$$

$$= [g\beta(T_w - T_e)x^2]^{1/2}\left[\frac{f'}{x}\left(\frac{1}{2}n + \frac{1}{2}\right) + f''\frac{\partial \eta}{\partial x}\right]$$

$$u\left(\frac{\partial u}{\partial x}\right)_y = [g\beta(T_w - T_e))]\left[f'^2\left(\frac{1}{2} + \frac{1}{2}n\right) + f'f''\frac{\partial \eta}{\partial x}x\right]$$

$$\frac{\partial u}{\partial y} = \frac{\partial u}{\partial \eta}\frac{\partial \eta}{\partial y} = [g\beta(T_w - T_e)x]^{1/2} f''\left[\frac{g\beta(T_w - T_e)}{\nu^2 x}\right]^{1/4}$$

$$v\frac{\partial u}{\partial y} = -[g\beta(T_w - T_e)]\left[\frac{1}{4}(3+n)ff'' + x\frac{\partial \eta}{\partial x}f'f''\right]$$

$$\nu\frac{\partial^2 u}{\partial y^2} = \nu[g\beta(T_w - T_e)x]^{1/2} f'''\left[\frac{g\beta(T_w - T_e)}{\nu^2 x}\right]^{1/2} = [g\beta(T_w - T_e)]f'''$$

$$\left(\frac{\partial T}{\partial x}\right)_y = \left(\frac{\partial T}{\partial x}\right)_\eta + \frac{\partial T}{\partial \eta}\frac{\partial \eta}{\partial x} = (T_w - T_e)\left[\frac{n}{x}\theta + \theta'\frac{\partial \eta}{\partial x}\right]$$

8. Buoyant Flows

$$u\frac{\partial T}{\partial x} = (T_w - T_e)[g\beta(T_w - T_e)x]^{1/2}\left[\frac{n}{x}\theta + \theta'\frac{\partial \eta}{\partial x}\right]f'$$

$$-v\frac{\partial T}{\partial y} = (T_w - T_e)[g\beta(T_w - T_e)x]^{1/2}\left[\frac{1}{4}(3+n)\frac{\theta'}{x}f + \theta'f'\frac{\partial \eta}{\partial x}\right]$$

$$\frac{\nu}{\Pr}\frac{\partial^2 T}{\partial y^2} = \frac{\theta''}{\Pr}(T_w - T_e)\left[\frac{g\beta(T_w - T_e)}{x}\right]^{1/2}.$$

Substituting the above equations into the momentum and energy equations and rearranging we get

$$f''' + \frac{3}{4}ff'' - \frac{1}{2}(f')^2 + \theta - \frac{1}{2}n\left[(f')^2 - \frac{1}{2}ff''\right] = 0 \tag{9}$$

$$\frac{\theta''}{\Pr} + \frac{3}{4}f\theta' + n\left(\frac{1}{4}f\theta' - f'\theta\right) = 0 \tag{10}$$

$$\eta = 0, \quad f = f' = 0, \quad \theta = 1; \quad \eta = \eta_e, \quad f'_e = 0, \quad \theta = 0$$

where

$$n = \frac{x}{T_w - T_e}\frac{d(T_w - T_e)}{dx}, \quad \theta(\eta) = \frac{T - T_e}{T_w - T_e}.$$

b. For specified wall flux, use

$$\eta = \left[\frac{g\beta(T_w - T_e)}{\nu^2 x}\right]^{1/4} y, \quad \psi = [g\beta(T_w - T_e)\nu^2 x^3]^{1/4}f(\eta)$$

$$T = T_e + T_e(\theta)\phi(x), \quad \phi(x) = -\frac{\dot{q}_w(x)}{k}\left[\frac{\nu^2 x}{g\beta(T_w - T_e)}\right]^{1/4}.$$

Under this transformation, the momentum equation is the same as the equation in part (a) with θ there defined by $\theta = (T - T_e)/(T_w - T_e)$. To obtain the energy equation for this case,

$$\left(\frac{\partial T}{\partial x}\right)_y = \left(\frac{\partial T}{\partial x}\right)_\eta + \frac{\partial T}{\partial \eta}\frac{\partial \eta}{\partial x} = T_e\theta\frac{\partial \phi}{\partial x} + T_e\theta'\phi\frac{\partial \eta}{\partial x}$$

$$u\left(\frac{\partial T}{\partial x}\right)_y = [g\beta(T_w - T_e)x]^{1/2}T_e\left[\theta\frac{\partial \phi}{\partial x}f' + \phi\theta'f'\frac{\partial \eta}{\partial x}\right]$$

$$\frac{\partial T}{\partial y} = \frac{\partial T}{\partial y}\frac{\partial \eta}{\partial y} = T_e\phi\theta'\left[\frac{g\beta(T_w - T_e)}{\nu^2 x}\right]^{1/4}$$

$$v\frac{\partial T}{\partial y} = -(T_e\phi\theta')[g\beta(T_w - T_e)x]^{1/2}\left[\left(\frac{1}{4}n + \frac{3}{4}\right)\frac{f}{x} + f'\frac{\partial \eta}{\partial x}\right]$$

$$\frac{\nu}{\Pr}\frac{\partial^2 T}{\partial y^2} = T_e\phi\theta''/\Pr\left[\frac{g\beta(T_w - T_e)}{x}\right]^{1/2}.$$

With the above relations, the energy equation becomes

$$\frac{T_e \phi \theta''}{\Pr} \left[\frac{g\beta(T_w - T_e)}{x} \right]^{1/2} = [g\beta(T_w - T_e)x]^{1/2}$$

$$(+T_e\phi) \left[\frac{\theta}{\phi} \frac{\partial \phi}{\partial x} f' - \left(\frac{1}{4}n + \frac{3}{4} \right) \frac{\theta' f}{x} \right].$$

Dividing both sides by $(-T_e\phi)[g\beta(T_w - T_e)/x]^{1/2}$ and rearranging we can write the energy equation and its boundary conditions as

$$\frac{\theta''}{\Pr} + \frac{3}{4}\theta' f + \frac{1}{4}nf\theta' - \hat{n}\theta f' = 0$$

$$\eta = 0, \quad \theta' = 1; \quad \eta = \eta_e, \quad \theta = 0$$

8.3 Problem

Obtain an expression for the magnitude and location of the maximum velocity of air ($\Pr = 0.72$) in a natural convection laminar boundary layer on a vertical plate with uniform wall temperature.

Solution

$$u = \frac{\partial \psi}{\partial y} = [g\beta(T_w - T_e)\nu^2 x^3]^{1/4} \frac{\partial f}{\partial \eta} \frac{\partial \eta}{\partial y}$$

$$= [g\beta(T_w - T_e)\nu^2 x^3]^{1/4} \left[\frac{g\beta(T_w - T_e)}{\nu^2 x} \right]^{1/4} f'$$

$$= [g\beta(T_w - T_e)L]^{1/2} \left(\frac{x}{L} \right)^{1/2} f'.$$

This implies that u is maximum when f' is maximum. From Fig. 8.2,

$$f'_{\max} \approx 0.55 \text{ for } \Pr = 0.72,$$

$$\therefore \quad u_{\max} = 0.55 \left(\frac{x}{L} \right)^{1/2} [g\beta(T_w - T_e)L]^{1/2}$$

8.4 Problem

Two uniformly heated vertical flat plates at $50\,^\circ\mathrm{C}$ are placed in air whose temperature is $15\,^\circ\mathrm{C}$. If the plates are 0.5 m high, what is the minimum spacing which will prevent merging of the natural convection boundary layers?

Solution

From Fig. 9.2, the edge of the boundary layer is $\eta_e = 5.5$. For

$$T_w = 50\,^\circ\mathrm{C}, \quad T_e = 15\,^\circ\mathrm{C}, \quad T_m = 1/2(T_w + T_e) = 32.5\,^\circ\mathrm{C},$$

$$\nu = 1.62 \times 10^{-5}\,\mathrm{m^2/s}, \quad \beta = 1/T_m \text{ and } x = 0.5\,\mathrm{m}.$$

8. Buoyant Flows

$$y_e = \left[\frac{\nu^2 x}{g\beta(T_w - T_e)}\right]^{1/4} \quad \eta_e = \left[\frac{(1.62 \times 10^{-5})^2 \times 0.5}{9.8 \times 35/(273 + 32.5)}\right]^{1/4} \times 5.5 = 0.018\,\text{m}.$$

To avoid the interference of the layers, the plates must be separated at least two times y_e, i.e. 0.036 m.

8.5 Problem

Repeat Example 8.1 when the fluid is (a) glycerin and (b) steam.

Solution

a. Glycerin: $\nu = 2.2 \times 10^{-4}\,\text{m}^2/\text{s}$, $\beta = 5.24 \times 10^{-4}$, Pr = 2.45 at 40°C

$$\text{Gr}_x = \frac{9.8 \times (0.25)^3 \times 5.24 \times 10^{-4}(65 - 15)}{(2.2 \times 10^{-4})^2} = 8.29 \times 10^4$$

$$-\theta'_w = 0.55, \quad f''_w = 0.77 \text{ for Pr} = 2.45$$

$$\therefore \quad c_f = \frac{2 \times 0.77}{(8.29 \times 10^4)^{1/2}} = 5.35 \times 10^{-3},$$

$$\text{Nu}_x = 0.55(8.29 \times 10^4)^{1/4} = 9.33$$

b. Water: From Table B-2

$$\beta = -\left(\frac{1}{\varrho}\frac{\partial \varrho}{\partial T}\right)_{T_f} = \frac{-1}{994.59} \times \frac{985.46 - 1000.52}{40} = 3.75 \times 10^{-4},$$

$$\nu = 0.658 \times 10^{-6}, \quad \text{Pr} = 4.34 \text{ at } 40\,°\text{C}$$

$$\text{Gr}_x = \frac{9.8 \times (0.25)^3 \times 3.75 \times 10^{-4}(65 - 15)}{(0.658 \times 10^{-6})^2} = 6.7 \times 10^9$$

$$f''_w = 0.70, \quad -\theta'_w = 0.66 \text{ for Pr} = 4.34 \text{ from Table 8.1}$$

$$\therefore \quad c_f = \frac{2.0 \times 0.7}{(6.7 \times 10^9)^{1/2}} = 1.71 \times 10^{-5}$$

$$\text{Nu}_x = 0.66 \times (6.7 \times 10^9)^{1/4} = 1.89 \times 10^2$$

8.6 Problem

Repeat Example 8.2 when the plate is *finite*. Discuss the comparison.

Solution

$$\bar{\dot{q}}_w = \frac{-k(T_w - T_e)}{L}0.28(\text{Gr}_L\text{Pr})^{1/3} = -\frac{k(T_w - T_e)}{L}\text{Gr}_L^{1/3} \times 0.25$$

$$\text{Nu}_x = \frac{\hat{h}x}{k} = \frac{\dot{q}_w}{(T_w - T_e)}\frac{x}{k}$$

$$\overline{\text{Nu}} = \int_0^L \text{Nu}_x\,dx = \frac{-\dot{q}_w}{(T_w - T_e)}\int_0^L \frac{x}{k}\,dx = \frac{L^2}{2k}\frac{\dot{q}_w}{(T_w - T_e)}$$

8.7 Problem

If the width of the plate in Example 8.3 is 0.5 m, calculate the total heat transfer from the plate.

Solution

The total heat flux, $Q_w = \int_0^L w \dot{q}_w \, dx$ is related to Nu by

$$\dot{q}_w = (T_w - T_e) \text{Nu}_x \frac{k}{x}$$

$$\therefore Q_w = w(T_w - T_e)k \int_0^L -\theta_w' \sqrt{R_x}/x = w(T_w - T_e)k(-\overline{\theta_w'}) 2\sqrt{R_L}$$

$$= 0.5(120 - 20) \times 0.03 \times 0.298 \times 2 \left(\frac{10 \times 1.5}{2.076 \times 10^{-5}}\right)^{1/2} = 760 \text{ W}$$

8.8 Problem

Consider a wall jet, with a slot height of 0.01 m, in which the slot and freestream velocities of air are equal and the freestream temperature is 300 °K.

(a) Determine the value of Richardson number for a freestream velocity of 0.1 m/s, and a uniform wall temperature of 320 °K.
(b) Determine the value of Richardson number for a freestream velocity of 1.0 m/s, and a uniform wall temperature of 320 °K.
(c) Determine the value of Richardson number for a freestream velocity of 1.0 m/s, and a uniform wall temperature of 1000 °K and consider the validity of the analysis for this case.

Solution

For an ideal gas,

$$\text{Ri} \equiv \frac{gL}{u_e^2} \frac{(T_w - T_e)}{T}$$

a. $\text{Ri} = \dfrac{9.81 \times 0.01}{0.01} \times \dfrac{20}{300} = 0.654$

b. $\text{Ri} = \dfrac{9.81 \times 0.01}{1} \times \dfrac{20}{300} = 6.54 \times 10^{-3}$

c. $\text{Ri} = \dfrac{9.81 \times 0.01}{1} \times \dfrac{700}{300} = 0.229$

Analysis is only valid for $\dfrac{\Delta T}{T} \to 0$.

8.9 PROBLEM

What do you expect to happen if the values of the freestream and wall temperature in example 8.4 are reversed?

SOLUTION

a. $\text{Ri} = \dfrac{9.81 \times 0.01}{0.01} \times \dfrac{200}{300} = 6.54$

b. $\text{Re} = \dfrac{0.1 \times 0.01}{1.6} \times 10^5 = 62.5$

$$z - z_0 = 10, \quad f_w'' \approx 20, \quad -\theta_w = 1.0$$

$$c_f/2 = \left(\dfrac{6.54}{62.5}\right)^{0.5} \dfrac{20}{(0.1/0.01)^{0.5}} = 2.03$$

$$\text{Nu} = 1 \times (10)^{0.5} \times \left(\dfrac{62.5}{6.54}\right)^{0.5} = 9.78$$

8.10 PROBLEM

Use Figs. 8.13–8.16 to access the values of $c_f/2$ and Nu for the conditions of Problem 8.1 but with a freestream velocity twice the slot velocity.

SOLUTION

$f_w'' \cong 20 \times 1.1, \quad -\theta_w' = 10 \times 5/6$

$$\therefore \quad c_f/2 = 2.03 \times 1.1 = 2.2,$$

$$\text{Nu} = 9.78 \times 5/6 \approx 8.15$$

9 Finite-Difference Solution of Boundary-Layer Equations: Internal Flows

9.1 Problem

Solve Eq. (E9.1.1) subject to the following boundary and initial conditions:

$$y = 0, \quad \frac{\partial T}{\partial y} = T, \quad y = 1, \quad \frac{\partial T}{\partial y} = -T, \qquad \text{(P9.1a)}$$

$$x = 0, \quad T = 1, \quad 0 \le y \le 1. \qquad \text{(P9.1b)}$$

Take $\alpha = 1$,

(a) an explicit method and employing central differences for the boundary conditions,

(b) an explicit method and employing a forward difference for the boundary condition at $y = 0$.

Compare the numerical results at $y = 0.3$ for $x = 0$, 0.20, 0.40, 0.80 and 1.0 obtained in each case with the analytical solution given by

$$T = 4 \sum_{n=1}^{\infty} \left[\frac{\sec \alpha_n}{(3 + 4\alpha_n^2)} e^{-4\alpha_n^2 x} \cos 2\alpha_n \left(y - \frac{1}{2} \right) \right], \quad 0 < y < 1 \qquad \text{(P9.2)}$$

where α_n are the positive roots of

$$\alpha \tan \alpha = \frac{1}{2}.$$

Take $\alpha = 1$, $\Delta x = 0.001$ and $\Delta y = 0.10$.

Solution

a. We use the computer program for Example 9.1 with changes to the boundary conditions at $y = 0$ and $y = 1$. We write Eq. (P9.1a) at $y = 0$ as

$$\frac{T_1^n - T_{-1}^n}{2\Delta y} = T_0^n. \qquad (1)$$

At $j = 0$, Eq. (E9.1.1) becomes

$$T_0^{n+1} = T_0^n + \frac{\Delta x}{(\Delta y)^2}(T_1^n - 2T_0^n + T_{-1}^n). \tag{2}$$

Eliminating T_{-1}^n between Eqs. (1) and (2) we obtain

$$T_0^{n+1} = T_0^n + \frac{\Delta x}{(\Delta y)^2}[T_1^n - (1 + \Delta y)T_0^n]. \tag{3}$$

At $y = 1$, or $j = J$, Eq. (P9.1a) becomes

$$T_J^{n+1} = T_J^n + \frac{\Delta x}{(\Delta y)^2}[T_{J+1}^n - 2T_J^n + T_{J-1}^n]. \tag{4}$$

In terms of central differences, the boundary conditions at $j = J$ can be written as

$$\frac{T_{J+1}^n - T_{J-1}^n}{2\Delta y} = -T_J^n \tag{5}$$

so that, similar to Eq. (4), Eq. (5) can be written as

$$T_J^{n+1} = T_J^n + 2\frac{\Delta x}{(\Delta y)^2}[T_{J-1}^n(1 + \Delta y)T_J^n]. \tag{6}$$

With the boundary conditions given by Eq. (3) at $j = 0$ and Eq. (6) at $j = J$, the computer program for this problem is similar to the one described for Example 9.1.

With $\Delta x = 0.001$ and $\Delta y = 0.10$, so that $r = 0.1$, the results at $y = 0.30$ obtained with the computer program given in the accompanying CD are presented below together with the analytical solutions of Eq. (P9.2).

Table 1. Comparision of numerical and analytical results at $y = 0.30$.

x	FDS	AS	Diff	%Error
.000	1.0000	1.0026	-.0026	-.0026
.200	.7351	.7348	.0003	.0004
.400	.5222	.5223	-.0001	-.0002
.600	.3709	.3712	-.0003	-.0008
.800	.2635	.2638	-.0004	-.0015
1.000	.1871	.1875	-.0004	-.0021

b. By following a procedure similar to that described for (a), we write the boundary condition at $y = 0$ as

$$T_0^{n+1} = (1 - \Delta y)T_1^{n+1}.$$

The rest, includung the boundary condition at $y = 1$, remains the same.

Table 2 presents results at $y = 0.30$ together with those obtained from the analytical solution. As can be seen, and as expected, the accuracy of numerical results are not as good as those obtained in (a).

Table 2. Comparison of numerical and analytical results at $y = 0.30$

r	FDS	AS	Diff	%Error
.000	1.0000	1.0026	-.0026	-.0026
.200	.7023	.7348	-.0325	-.0442
.400	.4849	.5223	-.0374	-.0716
.600	.3349	.3712	-.0363	-.0979
.800	.2313	.2638	-.0326	-.1235
1.000	.1597	.1875	-.0278	-.1483

9.2 PROBLEM

Repeat the above problem with the Crank–Nicolson method with central differences for the boundary conditions.

SOLUTION

As in Problem 9.1, we now make changes to the boundary conditions in the computer program used for Example 9.2. We make these changes by redefining b_1, c_1, and adding additional two rows to the a-matrix, Eq. (9.2.19b), at $j = 1$ and $j = J + 1$.

The new definitions of b_1 and c_1 are

$$b_1 = -(2 + \lambda_0) - 2\Delta y \qquad (1)$$
$$c_1 = 2$$

where

$$\lambda_0 = 2\frac{(\Delta y)^2}{\Delta x}. \qquad (2)$$

In addition

$$r_0 = -\lambda_0 T_0^n - 2[T_1^n - (1 + \Delta y)T_0]. \qquad (3)$$

At $j = J + 1$, we have

$$a_{J+1} = 2$$
$$b_{J+1} = -(2 + \lambda_0) - 2\Delta y \qquad (4)$$

$$r_{J+1} = -\lambda_0 T_{J+1}^n - 2[T_J^n - (1 + \Delta y)T_{J+1}^n]. \qquad (5)$$

The results show in Table 3 obtained with the revised computer program of Example 9.2 and given below are in excellent agreement with those obtained from Eq. (P9.2). The computer program is given in the accompanying CD.

Table 3. Comparison of numerical and analytical results at $y = 0.30$

x	FDS	AS	Diff	%Error
.000	1.0000	1.0026	-.0026	-.0026
.200	.7353	.7348	.0005	.0007
.400	.5225	.5223	.0002	.0004
.600	.3712	.3712	.0000	.0000
.800	.2638	.2638	-.0001	-.0003
1.000	.1874	.1875	-.0001	-.0006

9.3 Problem

Repeat Problem 9.1 with Keller's Box method and compare the solutions with the Crank–Nicolson method and Eq. (P9.2).

Solution

We use the computer program of Example 9.3 with minor changes to the boundary conditions at $y = 0$ and $y = 1$. Noting that the first row of A_0 and the last row of A_J in Eq. (9.2.29) denote boundary conditions, we set $\alpha_0 = 1$, $\alpha_1 = -1$ for wall boundary conditions and $\beta_0 = 1$, $\beta_1 = 1$. The rest of the computer program remains unchanged.

Table 4 presents the results together with those obtained from the analytical solution. Again, the agreement between numerical and analytical solutions is excellent. The computer program is given in the accompanying CD.

Table 4. Comparison of calculated and analytical results at $y = 0.30$

x	FDS	AS	Diff	%Error
.000	1.0000	1.0026	-.0026	-.0026
.200	.7362	.7348	.0014	.0019
.400	.2530	.5223	.0007	.0013
.600	.3715	.3712	.0003	.0008
.800	.2639	.2638	.0001	.0002
1.000	.1875	.1875	-.0001	-.0004

9.4 Problem

Solve Eq. (E9.1.1) subject to the following boundary and initial conditions:

$$y = 0, \quad T = 0, \quad y = 1, \quad T = 0, \quad \text{(P9.6a)}$$

$$x = 0, \quad T = \sin \pi y, \quad 0 \leq y \leq 1. \quad \text{(P9.6b)}$$

Use

(a) an explicit method and employing central differences for the boundary conditions,

(b) an explicit method and employing a forward difference for the boundary condition at $y = 0$.

Compare the numerical results obtained in each case with the analytical solution given by
$$T = e^{-\pi^2 x} \sin \pi y \tag{P9.7}$$
at $y = 0.5$ for $r = 0.1$ and 0.5 for values of $y = 0$, 0.2, 0.4, 0.6, 0.8 and 1.0.

SOLUTION

Except for the initial condition, the computer program for this problem is identical to the one used to solve Example 9.1. Numerical and analytical results are given in Tables 5 and 6 for both values of r and the computer program is given in the accompanying CD.

The numerical results in Table 5 for $r = 0.1$ are in excellent agreement with the analytical solutions. Those in Table 6 for $r = 0.5$, however and as expected, are not acceptable.

Table 5. Comparison of numerical and analytical results at $y = 0.30$, $r = 0.10$

x	FDS	AS	Diff	%Error
.000	.8090	.8090	.0000	.0000
.200	.1131	.1124	.0007	.0066
.400	.0158	.0156	.0002	.0132
.600	.0022	.0022	.0000	.0198
.800	.0003	.0003	.0000	.0265
1.000	.0000	.0000	.0000	.0332

Table 6. Comparison of numerical and analytical results at $y = 0.30$, $r = 0.5$

x	FDS	AS	Diff	%Error
.000	.8090	.8090	.0000	.0000
.200	.1258	.1124	.0134	.1194
.400	.0194	.0156	.0040	.2542
.600	.0030	.0022	.0009	.4053
.800	.0005	.0003	.0002	.5745
1.000	.0001	.0000	.0000	.7641

9.5 PROBLEM

Repeat the above problem with the Crank–Nicolson Method with central differences for the boundary conditions.

SOLUTION

Again the computer program for this problem, except for the initial condition, is the same as that used for Example 9.2. A comparison of numerical results with analytical solutions given in Table 7 indicates excellent agreement. The computer program is given in the accompanying CD.

Table 7. Comparison of numerical and analytical results at $y = 0.30$

x	FDS	AS	Diff	%Error
.000	.8090	.8090	.0000	.0000
.200	.1093	.1124	-.0031	-.0275
.400	.0147	.0156	-.0009	-.0590
.600	.0020	.0022	-.0002	-.0895
.800	.0003	.0003	.0000	-.1191
1.000	.0000	.0000	.0000	-.1476

9.6 PROBLEM

Repeat Problem 9.4 with Keller's Box method and compare the solutions with the Crank–Nicolson method and Eq. (P9.4).

SOLUTION

As before, the computer program for this problem, except for the initial condition, is same as that used for Example 9.3. A comparison between the two numerical solutions and the analytical solution given in Table 8 indicates very good agreement; this is of course expected since both methods are second-order accurate. The computer program is given in the accompanying CD.

Table 8. Comparison of numerical and analytical results at $y = 0.30$

x	FDS	AS	Diff	%Error
.000	.8090	.8090	.0000	.0000
.200	.1093	.1124	-.0031	-.0275
.400	.0147	.0156	-.0009	-.0590
.600	.0020	.0022	-.0002	-.0895
.800	.0003	.0003	.0000	-.1191
1.000	.0000	.0000	.0000	-.1476

11 Computer Programs and Their Applications to Momentum and Heat Transfer Problems

11.1 Problem

Sellars et al. have obtained analytical solutions for an incompressible laminar flow in a circular pipe with linear wall temperature [6]. They showed that for this case, the local Nusselt number is given by the expression

$$\mathrm{Nu} = \frac{\dfrac{1}{2} - 4\sum_{n=0}^{\infty}(G_n/\lambda_n^2)\exp\left(\dfrac{-\lambda_n^2 x^+}{\mathrm{Pe}}\right)}{\dfrac{88}{768} - 8\sum_{n=0}^{\infty}(G_n/\lambda_n^4)\exp\left(\dfrac{-\lambda_n^2 x^+}{\mathrm{Pe}}\right)} \quad \text{(P11.1)}$$

where λ_n, G_n are the same functions tabulated in Table 11.5. The solution of this equation for values of $x+ > 0.5\,\mathrm{Pe}$ also tends to a constant value given by Eq. (5.2.20), see Fig. 5.3.

(a) Modify the computer program given in Section 11.8 to obtain solutions of this problem numerically and compare the numerical results with those given by Eq. (P11.1) for $10^{-4}\,\mathrm{Pe} < x^+ < \infty$.

(b) For $x^+ \leq 0.001\,\mathrm{Pe}$, the solutions of Eq. (P11.1) can be approximated by

$$\mathrm{Nu} = \frac{2.035}{(x^+/\mathrm{Pe})^{1/3}}.$$

Compare your numerical results with this equation.

Solution

Assume a wall temperature variation given by $T_w = a + b\hat{x}$ and write it as $T_w - T_e = b\hat{x}$, since at $\hat{x} = 0$, $T_w = T_e$ and therefore $a = T_e$. With g now given by $g = (T_w - T)/(T_w - T_e)$, n is given in Eq. (5.3.7b) and is equal to $1/\hat{x}$. The definition of a_3 given by (5.3.7a) before the shear layers merge, becomes $a_3 = a_2$. After the shear layers merge, we solve Eqs. (5.3.8) and

(5.3.9) with the definitions of a_1 and a_2 and $\hat{n}$ which are the same as those before the merge, except that now a_3 is

$$a_3 = a_2 \frac{1}{\hat{x}} = \frac{a_2}{\hat{x}}.$$

In the computer code given in the accompanying CD, it is not necessary to modify the boundary conditions since the code is still for the specified wall temperature case. However, since there is no discontinuity at the boundary, it is not necessary to use the smoothing function. In this problem the expression for $(s_3)_j$ given by (9.3.4c) needs correction,

$$(s_3)_j = -\frac{1}{2}(a_3)_{j-1/2}^n - (a_2)_{j-1/2}^{n-1/2} a_n$$

as well as the expression for the Nusselt number which should be written in a more general form as $\mathrm{Nu} = 2g'_w/(g_m - g_w)$ rather than as $2g'_w/g_m$. See the computer program in the accompanying CD for further details. The numerical (solid) and analytical results (dashed) are given below.

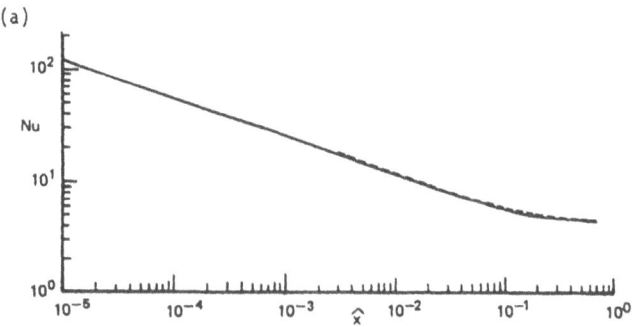

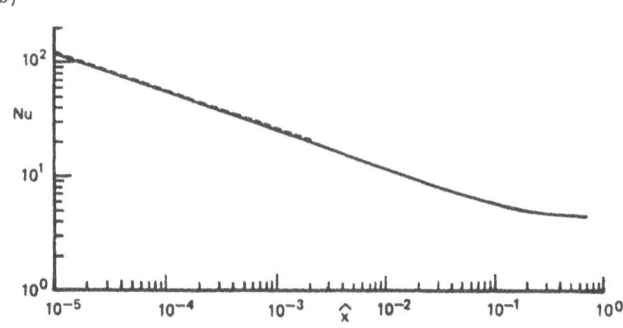

11.2 Problem

For certain boundary conditions the solutions of Eq. (5.1.3) can be obtained by analytical methods without resorting to numerical methods.

11. Computer Programs and Their Applications

(a) Show that for a two-dimensional slug flow, the solution of Eq. (5.1.3) subject to the boundary conditions

$$T(0, y) = T_e; \quad T(x, \pm h) = T_w;$$

and in the line of symmetry $\left(\dfrac{\partial T}{\partial y}\right)_{y=0} = 0$

can be written as

$$g(x^*, y^*) = \sum_{n=0}^{\infty} \frac{4}{\lambda_n} \sin \frac{\lambda_n}{2} \exp\left[-\frac{\lambda_n^2}{4} \frac{x^*}{G_2}\right] \cos \frac{\lambda_n}{2} y^*. \tag{P11.2}$$

Here $x^* = x/h$, $y^* = y/h$ and

$$g = \frac{T - T_w}{T_c - T_w}, \quad G_2 = R_h \Pr, \quad \lambda_n = (2n+1)\pi.$$

(b) Using the definition of local Nusselt number defined by Eq. (5.2.27) with d_c replaced by x, show that

$$\mathrm{Nu}_x = \frac{2x^* \sum_{n=0}^{\infty} \exp\left[-\dfrac{\lambda_n^2}{4} \dfrac{x^*}{G_2}\right]}{\int_0^1 g\, dy^*}. \tag{P11.3}$$

(c) Modify the computer program of Section 11.8 to obtain solutions to this problem, numerically and compare them with Eq. (P11.3).

SOLUTION

a. Let

$$g = \frac{T_w - T}{T_w - T_e}, \quad x^+ = \frac{x}{h}, \quad y^+ = \frac{y}{h}, \quad G_z = R_h \Pr = \frac{uh}{\nu}\Pr. \tag{1}$$

Then the solution of Eq. (5.1.3)

$$G_z \frac{\partial q}{\partial x^+} = \frac{\partial^2 g}{\partial y^{+2}} \tag{2}$$

is

$$g(x^+, y^+) = X(x^+) Y(y^+) \tag{3}$$

where X and Y satisfy

$$Y'' + A^2 Y = 0, \quad X' + \frac{A^2}{G_z} X = 0. \tag{4}$$

The solutions of (4) are

$$Y = a \cos Ay^* + b \sin Ay^*, \quad X = c \exp(-A^2/G_z x^+). \tag{5}$$

The constants are determined from $y^+ = 0$, $Y' = 0$, $y^+ = \pm 1$, $Y = 0$ to be $b = 0$, $A = (n + \frac{1}{2})\pi$ and, with $\lambda_n = (2n+1)\pi$,

$$Y = a \cos \frac{\lambda_n}{2} y^+, \quad X = c \exp\left(-\frac{\lambda_n^2}{4} \frac{x^+}{G_z}\right).$$

Since (2) is linear, any combination of X_1 and X_2 is also a solution. Thus

$$g = \sum_{n=0}^{\infty} a_n \exp\left(-\frac{\lambda_n^2}{4} \frac{x^+}{G_z}\right) \cos \frac{\lambda_n}{2} y^+$$

where a_n are determined from $g = 1.0$ at $x^+ = 0$, i.e.,

$$1 = \sum_{n=0}^{\infty} a_n \cos \frac{\lambda_n}{2} y^+$$

or

$$a_4 = \frac{4}{\lambda_n} \sin \frac{\lambda_n}{2} = \frac{4}{\lambda_n} \sin\left(n + \frac{1}{2}\right)\pi = (-1)^n \frac{4}{\lambda_n}$$

$$\therefore g(x^+, y^+) = \sum_{n=0}^{\infty} \frac{4}{\lambda_n} \sin \frac{\lambda_n}{2} \exp\left(-\frac{\lambda_n^2}{4} \frac{x^+}{G_z}\right) \cos \frac{\lambda_n}{2} y^+$$

$$= \sum_{n=0}^{\infty} (-1)^n \frac{4}{\lambda_n} \exp\left(-\frac{\lambda_n^2}{4} \frac{x^+}{G_z}\right) \cos \frac{\lambda_n}{2} y^+$$

b.

$$\dot{q}_w = -k \left(\frac{\partial T}{\partial y}\right)_w = -k(T_e - T_w) \frac{1}{h} g'(-1)$$

$$= -k(T_e - T_w)/h \sum_{n=0}^{\infty} 2 \exp\left(-\frac{\lambda_n^2}{4} \frac{x^+}{G_z}\right)$$

$$T_w - T_m = \int_0^1 (T_w - T) dy^+ = (T_w - T_e) \int_0^1 g \, dy^+$$

$$\therefore \mathrm{Nu}_x = \frac{\dot{q}_w}{T_w - T_m} \frac{x}{k} = \frac{2x^+ \sum_{n=0}^{\infty} \exp\left(-\frac{\lambda_n^2}{4} \frac{x^+}{G_z}\right)}{\int_0^1 g \, dy^+}$$

c. The results obtained with the computer program in the accompanying CD are given below.

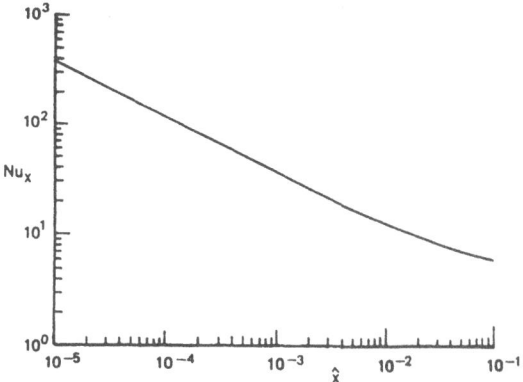

11.3 Problem

Use the computer program of Section 11.8, which provides a solution of Eq. (7.1.1) for a pipe with specified uniform wall temperature, to obtain solutions for $\Pr = 14.3$, $R_d = 10^5$. Plot

(a) the local Nusselt number variation, and
(b) the temperature profiles at $x/d = 2$, 4, 6 and 8.

Solution

Using the computer program of Section 11.8, we obtain (a) the variation of the local Nusselt number, and (b) the temperature profiles at $x/d = 2$, 4, 6 and 8 for $R_d = 10^5$, $\Pr = 14.3$.

(a)

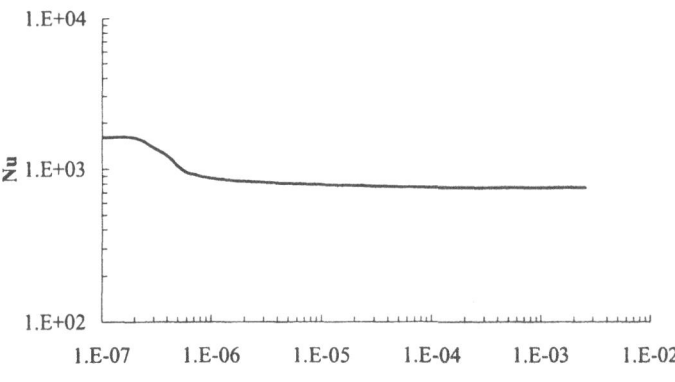

(b)

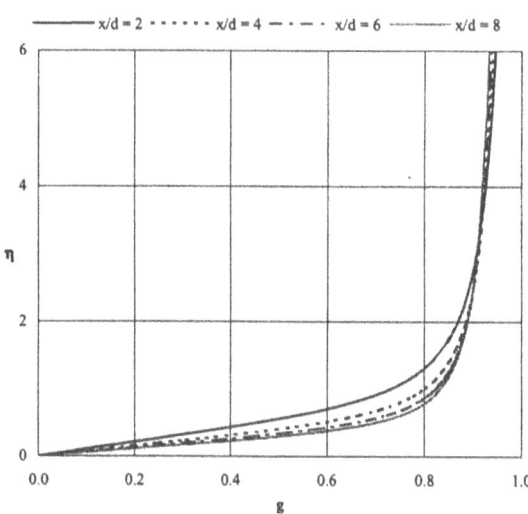

11.4 Problem

Repeat Problem 11.3 for $Pr = 0.02$.

Solution

Comparison of calculated results obtained from the computer program of Section 11.8 and those from Problem 7.10 are shown below.

(a)

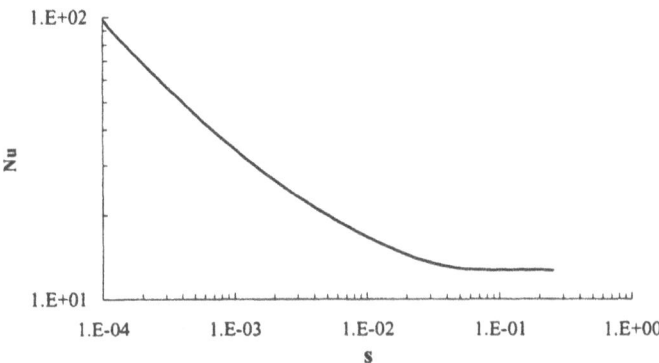

(b)

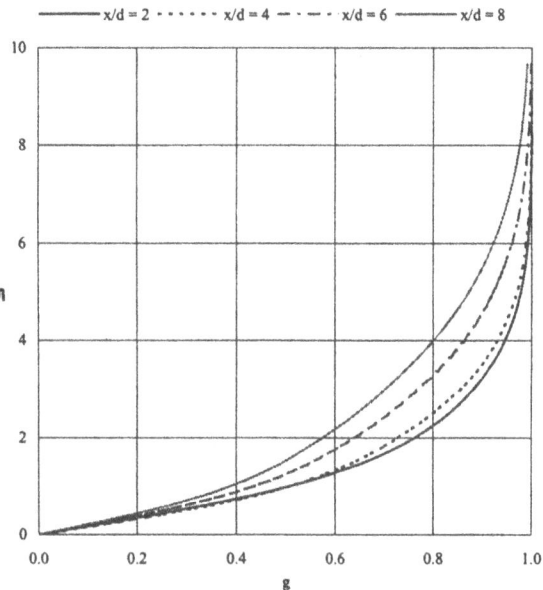

11.5 Problem

For an incompressible turbulent flow of gases in a circular pipe with constant heat flux, Reynolds, Swearingen and McEligot [7] show that the axial variation of Nusselt number is correlated by

$$\frac{\text{Nu}}{(\text{Nu})_\infty} = 1 + 0.8(1 + 70{,}000 R_d^{-1.5})\left(\frac{x}{d}\right)^{-1} \qquad \text{(P11.4)}$$

to within ± 5 percent for $x/d \geq 2$ and for a range of Reynolds numbers from 3,000 to 50,000. Here Nu is given by Eq. (5.2.18b) and $(\text{Nu})_\infty$ denotes the Nusselt number corresponding to fully developed conditions obtained from the empirical formula of Dittus and Boelter,

$$(\text{Nu})_\infty = 0.021 R_d^{0.8} \text{Pr}^{0.4}. \qquad \text{(P11.5)}$$

Using the computer program of Section 11.8, compare the numerical results with those given by Eqs. (P11.4) and (P11.5). Take $\text{Pr} = 0.72$.

Solution

Comparison of calculated results with those given by Eqs. (P11.4, dashed line) and (P11.5, solid line) for constant heat flux and $R_d = 10^4$ are shown below.

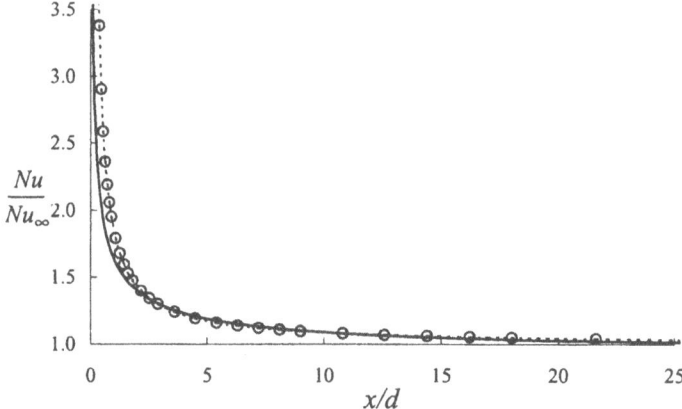

11.6 Problem

A gas turbine blade is based on an NACA 2412 airfoil and has the geometry and external velocity distribution given below. It operates at the following conditions:

$$\begin{aligned}
\text{total temperature,} \quad & T_0 = 850\,°\text{C} \\
\text{kinematic viscosity,} \quad & \nu = 6 \times 10^{-5}\,\text{m}^2\,\text{s}^{-1}. \\
\text{freestream velocity,} \quad & u_\infty = 200\,\text{m}\,\text{s}^{-1}.
\end{aligned}$$

We wish to determine the amount of cooling $(\text{kW}\,\text{m}^{-1})$ needed to maintain the surface temperature at $550\,°\text{C}$ from the leading-edge to the transition point, then, decreasing linearly to $450\,°\text{C}$ at the trailing edge. We are also interested in determining the local skin-friction distribution and displacement thickness distributions on the upper and lower surfaces of the blade. Suggested procedure is:

1. Determine momentum and heat transfer for laminar flow.
2. Compute transition location.
3. Determine momentum and heat transfer for turbulent flow.

Take the chord length to be $0.1\,\text{m}$.

Coordinates and external (inviscid) velocity distribution of NACA 2412 airfoil for a zero degree angle of attack

x/c	y/c	u_e/u_∞
0.975017	−0.002574	−0.923056
0.925000	−0.006500	−0.951147
0.850000	−0.011600	−0.967792
0.750002	−0.018235	−0.989543
0.650002	−0.024535	−1.006718
0.550005	−0.030586	−1.026400
0.450007	−0.035862	−1.046163
0.350006	−0.039787	−1.064401
0.275001	−0.041775	−1.084731
0.225000	−0.042390	−1.103499
0.174994	−0.041869	−1.122426
0.124977	−0.039573	−1.138186
0.087484	−0.036186	−1.154834
0.062454	−0.032608	−1.166511
0.037354	−0.026895	−1.159470
0.018586	−0.019930	−1.124105
0.004651	−0.009461	−0.729720
0.003899	0.012117	0.603138
0.018541	0.026010	1.075167
0.037283	0.036076	1.156801
0.062418	0.045696	1.198170
0.087457	0.053112	1.219485
0.124906	0.061678	1.241882
0.174955	0.069695	1.249947
0.224978	0.074922	1.249534
0.274990	0.077978	1.244678
0.350006	0.079106	1.227716
0.450027	0.075688	1.194753
0.550034	0.068384	1.162275
0.650039	0.058038	1.129654
0.750043	0.044950	1.091903
0.850053	0.029465	1.046916
0.925026	0.016242	1.000793
0.975055	0.005940	0.923055

SOLUTION

The input geometry and the external velocity distribution for this problem can be read into the boundary layer codes for momentum and heat transfer. Here, for convenience, we follow a slightly different approach as described in steps 1–3 in the accompanying CD. There we also discuss and describe our procedure for calculating momentum and heat transfer for both laminar and turbulent flows, including the prediction of the transition location.

11.7 Problem

Show that the computer program for Head's method presented in Section 11.9 for two-dimensional flows can be extended to axisymmetric turbulent boundary layers.

Hint: Note that:

$$B(2) = u_e r_0 \theta H_1$$
$$C(1) = \frac{c_f}{2} - (H+2)\frac{\theta}{u_e}\frac{du_e}{dx} - \frac{\theta}{r_0}\frac{dr_0}{dx}$$
$$C(2) = r_0 u_e F.$$

In addition it is necessary to read in $r_0(x)/L$ and compute dr_0/dx or $d/d(x/L)(r_0/L)$. The latter can be done with a three-point Lagrange-interpolation formula (see subsection 11.2.2).

SOLUTION

Laminar flow calculations are performed with Thwaites' method and those for turbulent flow with Head's method. In the axisymmetric case, the values of $x(I)$ are replaced by the surface distance, and the values of $r_0(x)/L$ (or $y(x)/L$) are input to the computer program. The necessary changes to the computer program presented in Section 11.9 for axisymmetric flow are as follows:

a. Modify common statement to include the arrays R(41) and DRDX(41) where R(I) is the array containing $r_0/L(x/L)$ and DRDX(I) is the array containing $[d(r_0/L)]/[d(x/L)]$.

b. Use the three-point Lagrange interpolation formula to calculate $[d(r_0/L)]/[d(x/L)]$:

    DRDX(1)   = RGNG[X(1),X(2),X(3),R(1),R(2),R(3),X(1)]
    DRDX(NXT) = RGNG[X(NXT-2),X(NXT-1),X(NXT),R(NXT-2),R(NXT-1),R(NXT),X(NXT)]

c. Modify DO loop 10 as:

    DO 10 I = 2, NXTM1,
    DUEDX = ...,
    DRDX(I) = RGNG[X(I-1),X(I),X(I+1),R(I-1),R(I),R(I+1),X(I)]

d. Modify the expression S(1) as:

    S(1) = UE(1)IT(1)*R(1)*HDFH1(-H(1))

e. In subroutine STNDRD, modify

    C(1) = -(HB+2.0)*B(1)/UI*UP+CF02-B(1)/R(1)*UP*DRDX(I)
    H1 = B(2)/B(1)/UI/R(1)
    C(2) = UI*0.0306/(H1-3.0)**0.6169*R(I)

Note that X is the surface distance and the derivatives must be evaluated with respect to their surface coordinates.

Appendix

A. Conversion Factors

1. **Acceleration**
   $1\,\text{ft}\,\text{s}^{-2} = 0.3048\,\text{m}\,\text{s}^{-2}$
   $1\,\text{m}\,\text{s}^{-2} = 3.2808\,\text{ft}\,\text{s}^{-2}$

2. **Area**
   $1\,\text{in}^2 = 6.4516\,\text{cm}^2$
   $1\,\text{ft}^2 = 0.0929\,\text{m}^2$
   $1\,\text{m}^2 = 10.764\,\text{ft}^2$

3. **Density**
   $1\,\text{lb}\,\text{in}^{-3} = 27.680\,\text{g}\,\text{cm}^{-3}$
   $1\,\text{lb}\,\text{ft}^{-3} = 16.019\,\text{kg}\,\text{m}^{-3}$
   $1\,\text{kg}\,\text{m}^{-3} = 0.06243\,\text{lb}\,\text{ft}^{-3}$
   $1\,\text{slug}\,\text{ft}^{-3} = 515.38\,\text{kg}\,\text{m}^{-3}$

4. **Diffusivity (heat, mass, momentum)**
   $1\,\text{ft}^2\,\text{s}^{-1} = 0.0929\,\text{m}^2\,\text{s}^{-1}$
   $1\,\text{ft}^2\,\text{h}^{-1} = 0.2581 \times 10^{-4}\,\text{m}^2\,\text{s}^{-1}$
   $1\,\text{m}^2\,\text{s}^{-1} = 10.7639\,\text{ft}^2\,\text{s}^{-1}$
   $1\,\text{cm}^2\,\text{s}^{-1} = 3.8745\,\text{ft}^2\,\text{h}^{-1}$

5. **Energy, heat, power**
   $1\,\text{J} = 1\,\text{W}\,\text{s} = 1\,\text{N}\,\text{m}$
   $1\,\text{J} = 10^7\,\text{erg}$
   $1\,\text{Btu} = 1055.04\,\text{J}$
   $1\,\text{Btu} = 1055.04\,\text{W}\,\text{s}$
   $1\,\text{Btu} = 252\,\text{cal}$
   $1\,\text{Btu} = 778.161\,\text{ft}\,\text{lb}_f$
   $1\,\text{Btu}\,\text{h}^{-1} = 0.2931\,\text{W}$
   $1\,\text{Btu}\,\text{h}^{-1} = 3.93 \times 10^{-4}\,\text{hp}$
   $1\,\text{cal} = 4.1868\,\text{J}$ (or $\text{W}\,\text{s}$ or $\text{N}\,\text{m}$)
   $1\,\text{cal} = 3.968 \times 10^{-3}\,\text{Btu}$
   $1\,\text{hp} = 550\,\text{ft}\,\text{lb}_f\,\text{s}^{-1}$
   $1\,\text{hp} = 745.7\,\text{W} = 745.7\,\text{N}\,\text{m}\,\text{s}^{-1}$
   $1\,\text{Wh} = 3.413\,\text{Btu}$

6. **Heat capacity, heat per unit mass, specific heat**
   $1\,\text{Btu}\,\text{h}^{-1}\,°\text{F}^{-1} = 0.5274\,\text{W}\,°\text{C}^{-1}$
   $1\,\text{W}\,°\text{C}^{-1} = 1.8961\,\text{Btu}\,\text{h}^{-1}\,°\text{F}^{-1}$
   $1\,\text{Btu}\,\text{lb}^{-1} = 2325.9\,\text{J}\,\text{kg}^{-1}$
   $1\,\text{Btu}\,\text{lb}^{-1}\,°\text{F}^{-1} = 4.18669\,\text{kJ}\,\text{kg}^{-1}\,°\text{C}^{-1}$ (or $\text{J}\,\text{g}^{-1}\,°\text{C}^{-1}$)
   $1\,\text{Btu}\,\text{lb}^{-1}\,°\text{F}^{-1} = 1\,\text{cal}\,\text{g}^{-1}\,°\text{C}^{-1} = 1\,\text{kcal}\,\text{kg}^{-1}\,°\text{C}^{-1}$

7. **Heat flux**
   $1\,\text{Btu}\,\text{h}^{-1}\,\text{ft}^{-2} = 3.1537 \times 10^{-3}\,\text{kW}\,\text{m}^{-2}$
   $1\,\text{W}\,\text{m}^{-2} = 0.31709\,\text{Btu}\,\text{h}^{-1}\,\text{ft}^{-2}$

8. **Heat-generation rate**
   $1\,\text{Btu}\,\text{h}^{-1}\,\text{ft}^{-3} = 10.35\,\text{W}\,\text{m}^{-3}$
   $1\,\text{W}\,\text{m}^{-3} = 0.0966\,\text{Btu}\,\text{h}^{-1}\,\text{ft}^{-3}$

9. **Heat-transfer coefficient**
   $1\,\text{Btu}\,\text{h}^{-1}\,\text{ft}^{-2}\,°\text{F}^{-1} = 5.677\,\text{W}\,\text{m}^{-2}\,°\text{C}^{-1}$
   $1\,\text{W}\,\text{m}^{-2}\,°\text{C}^{-1} = 0.1761\,\text{Btu}\,\text{h}^{-1}\,\text{ft}^{-2}\,°\text{F}^{-1}$
   $1\,\text{Btu}\,\text{h}^{-1}\,\text{ft}^{-2}\,°\text{F}^{-1} = 4.882\,\text{kcal}\,\text{h}^{-1}\,\text{m}^{-2}\,°\text{C}^{-1}$

10. **Length**
    $1\,\text{in} = 2.54\,\text{cm}$
    $1\,\text{in} = 2.54 \times 10^{-2}\,\text{m}$
    $1\,\text{ft} = 0.3048\,\text{m}$
    $1\,\text{m} = 3.2808\,\text{ft}$
    $1\,\text{mile} = 1609.34\,\text{m}$
    $1\,\text{mile} = 5280\,\text{ft}$

11. Mass

1 oz = 28.35 g
1 lb = 16 oz
1 lb = 453.6 g 1 lb = 0.4536 kg
1 kg = 2.2046 lb
1 g = 15.432 grains
1 slug = 32.1739 lb

12. Mass flux

$1 \, \text{lb ft}^{-2} \, \text{h}^{-1} = 1.3563 \times 10^{-3} \, \text{kg m}^{-2} \, \text{s}^{-1}$
$1 \, \text{lb ft}^{-2} \, \text{s}^{-1} = 4.882 \, \text{kg m}^{-2} \, \text{s}^{-1}$
$1 \, \text{kg m}^{-2} \, \text{s}^{-1} = 737.3 \, \text{lb ft}^{-2} \, \text{h}^{-1}$
$1 \, \text{kg m}^{-2} \, \text{s}^{-1} = 0.2048 \, \text{lb ft}^{-2} \, \text{s}^{-1}$

13. Pressure, force

$1 \, \text{N} = 1 \, \text{kg m s}^{-2}$
$1 \, \text{N} = 0.2248 \, \text{lb}_f$
$1 \, \text{N} = 7.2333 \, \text{poundals}$
$1 \, \text{N} = 10^5 \, \text{dyn}$
$1 \, \text{N m}^{-2} = 1 \, \text{Pa}$
$1 \, \text{lb}_f = 32.174 \, \text{ft lb s}^{-2}$
$1 \, \text{lb}_f = 4.4482 \, \text{N}$
$1 \, \text{lb}_f = 32.1739 \, \text{poundals}$
$1 \, \text{lb}_f \, \text{in}^{-2} \equiv (1 \, \text{psi}) = 6894.76 \, \text{Pa}$
$1 \, \text{lb}_f \, \text{ft}^{-2} = 47.880 \, \text{Pa}$
$1 \, \text{bar} = 10^5 \, \text{Pa}$
$1 \, \text{atm} = 14.696 \, \text{lb}_f \, \text{in}^{-2}$
$1 \, \text{atm} = 2116.2 \, \text{lb}_f \, \text{ft}^{-2}$
$1 \, \text{atm} = 1.0132 \times 10^5 \, \text{Pa}$

14. Specific heat

$1 \, \text{Btu lb}^{-1} \, °\text{F}^{-1} = 1 \, \text{kcal kg}^{-1} \, °\text{C}^{-1}$
$\quad = 1 \, \text{cal g}^{-1} \, °\text{C}^{-1}$
$1 \, \text{Btu lb}^{-1} \, °\text{F}^{-1} = 4.18669 \, \text{J g}^{-1} \, °\text{K}^{-1}$
$\quad (\text{or W s g}^{-1} \, °\text{C}^{-1})$
$1 \, \text{J g}^{-1} \, °\text{C}^{-1} = 0.23885 \, \text{Btu lb}^{-1} \, °\text{F}^{-1}$
$\quad (\text{cal g}^{-1} \, °\text{C}^{-1} \text{ or kcal kg}^{-1} \, °\text{C}^{-1})$

15. Speed

$1 \, \text{ft s}^{-1} = 0.3048 \, \text{m s}^{-1}$
$1 \, \text{m s}^{-1} = 3.2808 \, \text{ft s}^{-1}$
$1 \, \text{mile h}^{-1} = 1.4667 \, \text{ft s}^{-1}$
$1 \, \text{mile h}^{-1} = 0.44704 \, \text{m s}^{-1}$

16. Temperature

$1 \, °\text{K} = 1.8 \, °\text{R}$
$\text{T} \, (°\text{F}) = 1.8(°\text{K} - 273) + 32$
$\text{T} \, (°\text{K}) = \frac{1}{1.8}(°\text{F} - 32) + 273$
$\text{T} \, (°\text{C}) = \frac{1}{1.8}(°\text{R} - 492)$

17. Thermal conductivity

$1 \, \text{Btu h}^{-1} \, \text{ft}^{-1} \, °\text{F}^{-1}$
$\quad = 1.7303 \, \text{W m}^{-1} \, °\text{C}^{-1}$
$1 \, \text{Btu h}^{-1} \, \text{ft}^{-1} \, °\text{F}^{-1}$
$\quad = 0.4132 \, \text{cal s}^{-1} \, \text{m}^{-1} \, °\text{C}^{-1}$
$1 \, \text{W m}^{-1} \, °\text{C}^{-1}$
$\quad = 0.5779 \, \text{Btu h}^{-1} \, \text{ft}^{-1} \, °\text{F}^{-1}$

18. Thermal resistance

$1 \, \text{h}^{-1} \, °\text{F}^{-1} \, \text{Btu}^{-1} = 1.896 \, °\text{C W}^{-1}$
$1 \, °\text{C W}^{-1} = 0.528 \, \text{h} \, °\text{F Btu}^{-1}$

19. Viscosity

$1 \, \text{poise} = 1 \, \text{g cm}^{-1} \, \text{s}^{-1}$
$1 \, \text{poise} = 10^2 \, \text{centipoise}$
$1 \, \text{poise} = 241.9 \, \text{lb ft}^{-1} \, \text{h}^{-1}$
$1 \, \text{lb ft}^{-1} \, \text{s}^{-1} = 1.4882 \, \text{kg m}^{-1} \, \text{s}^{-1}$
$1 \, \text{lb ft}^{-1} \, \text{s}^{-1} = 14.882 \, \text{poises}$
$1 \, \text{lb ft}^{-1} \, \text{h}^{-1} = 0.4134 \times 10^{-3} \, \text{kg m}^{-1} \, \text{s}^{-1}$
$1 \, \text{lb ft}^{-1} \, \text{h}^{-1} = 0.4134 \times 10^{-2} \, \text{poise}$

20. Volume

$1 \, \text{in}^3 = 16.387 \, \text{cm}^3$
$1 \, \text{cm}^3 = 0.06102 \, \text{in}^3$
$1 \, \text{oz (U.S. fluid)} = 29.573 \, \text{cm}^3$
$1 \, \text{ft}^3 = 0.0283168 \, \text{m}^3$
$1 \, \text{ft}^3 = 28.3168 \, \text{liters}$
$1 \, \text{ft}^3 = 7.4805 \, \text{gal (U.S.)}$
$1 \, \text{m}^3 = 35.315 \, \text{ft}^3$
$1 \, \text{gal (U.S.)} = 3.7854 \, \text{liters}$
$1 \, \text{gal (U.S.)} = 3.7854 \times 10^{-3} \, \text{m}^3$
$1 \, \text{gal (U.S.)} = 0.13368 \, \text{ft}^3$

A. Conversion Factors

Constants

g_c = gravitational acceleration conversion factor
$\quad$ = $32.1739 \, \text{ft lb lb}_f^{-1} \, \text{s}^{-2}$
$\quad$ = $4.1697 \times 10^8 \, \text{ft lb lb}_f^{-1} \, \text{h}^{-2}$
$\quad$ = $1 \, \text{g cm dyn}^{-1} \, \text{s}^{-2}$
$\quad$ = $1 \, \text{kg m N}^{-1} \, \text{s}^{-2}$
$\quad$ = $1 \, \text{lb ft poundal}^{-1} \, \text{s}^{-2}$
$\quad$ = $1 \, \text{slug ft lb}_f^{-1} \, \text{s}^{-2}$

J = mechanical equivalent of heat = $778.16 \, \text{ft lb}_f \, \text{Btu}^{-1}$

R = gas constant = $1544 \, \text{ft lb}_f \, \text{lb}^{-1} \, \text{mol}^{-1} \, °\text{R}^{-1}$
$\quad$ = $8.314 \, \text{N m g}^{-1} \, \text{mol}^{-1} \, °\text{K}^{-1}$
$\quad$ = $1.987 \, \text{cal g}^{-1} \, \text{mol}^{-1} \, °\text{K}^{-1}$

B. Physical Properties of Gases, Liquids, Liquid Metals, and Metals

Table B.1. Physical properties of gases at atmospheric pressure

T (°K)	ϱ (kg m^{-3})	c_p (kJ kg^{-1} °K^{-1})	μ kg m^{-1} s^{-1}	ν (m^2 s^{-1} ×10^6)	k (W m^{-1} °K^{-1})	κ (m^2 s^{-1} ×10^4)	Pr
Air							
100	3.6010	1.0266	0.6924×10^{-5}	1.923	0.009246	0.02501	0.770
150	2.3675	1.0099	1.0283	4.343	0.013735	0.05745	0.753
200	1.7684	1.0061	1.3289	7.490	0.01809	0.10165	0.739
250	1.4128	1.0053	1.488	9.49	0.02227	0.13161	0.722
300	1.1774	1.0057	1.983	15.68	0.02624	0.22160	0.708
350	0.9980	1.0090	2.075	20.76	0.03003	0.2983	0.697
400	0.8826	1.0140	2.286	25.90	0.03365	0.3760	0.689
450	0.7833	1.0207	2.484	28.86	0.03707	0.4222	0.683
500	0.7048	1.0295	2.671	37.90	0.04038	0.5564	0.680
550	0.6423	1.0392	2.848	44.34	0.04360	0.6532	0.680
600	0.5870	1.0551	3.018	51.34	0.04659	0.7512	0.680
650	0.5430	1.0635	3.177	58.51	0.04953	0.8578	0.682
700	0.5030	1.0752	3.332	66.25	0.05230	0.9672	0.684
750	0.4709	1.0856	3.481	73.91	0.05509	1.0774	0.686
800	0.4405	1.0978	3.625	82.29	0.05779	1.1951	0.689
850	0.4149	1.1095	3.765	90.75	0.06028	1.3097	0.692
900	0.3925	1.1212	3.899	99.3	0.06279	1.4271	0.696
950	0.3716	1.1321	4.023	108.2	0.06525	1.5510	0.699
1000	0.3524	1.1417	4.152	117.8	0.06752	1.6779	0.702
1100	0.3204	1.160	4.44	138.6	0.0732	1.969	0.704
1200	0.2947	1.179	4.69	159.1	0.0782	2.251	0.707
1300	0.2707	1.197	4.93	182.1	0.0837	2.583	0.705
1400	0.2515	1.214	5.17	205.5	0.0891	2.920	0.705
1500	0.2355	1.230	5.40	229.1	0.0946	3.262	0.705
1600	0.2211	1.248	5.63	254.5	0.100	3.609	0.705
1700	0.2082	1.267	5.85	280.5	0.105	3.977	0.705
1800	0.1970	1.287	6.07	308.1	0.111	4.379	0.704
1900	0.1858	1.309	6.29	338.5	0.117	4.811	0.704
2000	0.1762	1.338	6.50	369.0	0.124	5.260	0.702
2100	0.1682	1.372	6.72	399.6	0.131	5.715	0.700
2200	0.1602	1.419	6.93	432.6	0.139	6.120	0.707
2300	0.1538	1.482	7.14	464.0	0.149	6.540	0.710
2400	0.1458	1.574	7.35	504.0	0.161	7.020	0.718
2500	0.1394	1.688	7.57	543.5	0.175	7.441	0.730

Table B.1 (continued)

T ($°K$)	ϱ ($kg\,m^{-3}$)	c_p ($kJ\,kg^{-1}\,°K^{-1}$)	μ ($kg\,m^{-1}\,s^{-1}$)	ν ($m^2\,s^{-1}\times 10^6$)	k ($W\,m^{-1}\,°K^{-1}$)	κ ($m^2\,s^{-1}\times 10^4$)	Pr
Helium							
3		5.200	8.42×10^{-7}		0.0106		
33	1.4657	5.200	50.2	3.42	0.0353	0.04625	0.74
144	3.3799	5.200	125.5	37.11	0.0928	0.5275	0.70
200	0.2435	5.200	156.6	64.38	0.1177	0.9288	0.694
255	0.1906	5.200	181.7	95.50	0.1357	1.3675	0.70
366	0.13280	5.200	230.5	173.6	0.1691	2.449	0.71
477	0.10204	5.200	275.0	269.3	0.197	3.716	0.72
589	0.08282	5.200	311.3	375.8	0.225	5.215	0.72
700	0.07032	5.200	347.5	494.2	0.251	6.661	0.72
800	0.06023	5.200	381.7	634.1	0.275	8.774	0.72
900	0.05286	5.200	413.6	781.3	0.298	10.834	0.72
Carbon dioxide							
220	2.4733	0.783	11.105×10^6	4.490	0.010805	0.05920	0.818
250	2.1657	0.804	12.590	5.813	0.012884	0.07401	0.793
300	1.7973	0.871	14.958	8.321	0.016572	0.10588	0.770
350	1.5362	0.900	17.205	11.19	0.02047	0.14808	0.755
400	1.3424	0.942	19.32	14.39	0.02461	0.19463	0.738
450	1.1918	0.980	21.34	17.90	0.02897	0.24813	0.721
500	1.0732	1.013	23.26	21.67	0.03352	0.3084	0.702
550	0.9739	1.047	25.08	25.74	0.03821	0.3750	0.685
600	0.8938	1.076	26.83	30.02	0.04311	0.4483	0.668
Carbon monoxide							
220	1.55363	1.0429	13.832×10^{-6}	8.903	0.01906	0.11760	0.758
250	0.8410	1.0425	15.40	11.28	0.02144	0.15063	0.750
300	1.13876	1.0421	17.843	15.67	0.02525	0.21280	0.737
350	0.97425	1.0434	20.09	20.62	0.02883	0.2836	0.728
400	0.85363	1.0484	22.19	25.99	0.03226	0.3605	0.722
450	0.75848	1.0551	24.18	31.88	0.0436	0.4439	0.718
500	0.68223	1.0635	26.06	38.19	0.03863	0.5324	0.718
550	0.62024	1.0756	27.89	44.97	0.04162	0.6240	0.721
600	0.56850	1.0877	29.60	52.06	0.04446	0.7190	0.724
Ammonia, NH_3							
220	0.3828	2.198	7.255×10^{-6}	19.0	0.0171	0.2054	0.93
273	0.7929	2.177	9.353	11.8	0.0220	0.1308	0.90
323	0.6487	2.177	11.035	17.0	0.0270	0.1920	0.88
373	0.5590	2.236	12.886	23.0	0.0327	0.2619	0.87
423	0.4934	2.315	14.672	29.7	0.0391	0.3432	0.87
473	0.4405	2.395	16.49	37.4	0.0467	0.4421	0.84

Table B.1 (continued)

T (°K)	ϱ (kg m^{-3})	c_p (kJ kg^{-1} °K^{-1})	μ kg m^{-1} s^{-1}	ν (m^2 s^{-1} ×10^6)	k (W m^{-1} °K^{-1})	κ (m^2 s^{-1} ×10^4)	Pr
Steam (H$_2$O vapor)							
380	0.5863	2.060	12.71×10^{-6}	21.6	0.0246	0.2036	1.060
400	0.5542	2.014	13.44	24.2	0.0261	0.2338	1.040
450	0.4902	1.980	15.25	31.1	0.0299	0.307	1.010
500	0.4405	1.985	17.04	38.6	0.0339	0.387	0.996
550	0.4005	1.997	18.84	47.0	0.0379	0.475	0.991
600	0.3652	2.026	20.67	56.6	0.0422	0.573	0.986
650	0.3380	2.056	22.47	64.4	0.0464	0.666	0.995
700	0.3140	2.085	24.26	77.2	0.0505	0.772	1.000
750	0.2931	2.119	26.04	88.8	0.0549	0.883	1.005
800	0.2739	2.152	27.86	102.0	0.0592	1.001	1.010
850	0.2579	2.186	29.69	115.2	0.0637	1.130	1.019
Hydrogen							
30	0.84722	10.840	1.606×10^{-6}	1.895	0.0228	0.02493	0.759
50	0.50955	10.501	2.516	4.880	0.0362	0.0676	0.721
100	0.24572	11.229	4.212	17.14	0.0665	0.2408	0.712
150	0.16371	12.602	5.595	34.18	0.0981	0.475	0.718
200	0.12270	13.540	6.813	55.53	0.1282	0.772	0.719
250	0.09819	14.059	7.919	80.64	0.1561	1.130	0.713
300	0.08185	14.314	8.963	109.5	0.182	1.554	0.706
350	0.07016	14.436	9.954	141.9	0.206	2.031	0.697
400	0.06135	14.491	10.864	177.1	0.228	2.568	0.690
450	0.05462	14.499	11.779	215.6	0.251	3.164	0.682
500	0.04918	14.507	12.636	257.0	0.272	3.817	0.675
550	0.04469	14.532	13.475	301.6	0.292	4.516	0.668
600	0.04085	14.537	14.285	349.7	0.315	5.306	0.664
700	0.03492	14.574	15.89	455.1	0.351	6.903	0.659
800	0.03060	14.675	17.40	569	0.384	8.563	0.664
900	0.02723	14.821	18.78	690	0.412	10.217	0.676
1000	0.02451	14.968	20.16	822	0.440	11.997	0.686
1100	0.02227	15.165	21.46	965	0.464	13.726	0.703
1200	0.02050	15.366	22.75	1107	0.488	15.484	0.715
1300	0.01890	15.575	24.08	1273	0.512	17.394	0.733
1333	0.01842	15.638	24.44	1328	0.519	18.013	0.736

Table B.1 (continued)

T ($^\circ$K)	ϱ (kg m^{-3})	c_p (kJ kg^{-1} $^\circ$K^{-1})	μ (kg m^{-1} s^{-1})	ν (m^2 s^{-1} $\times 10^6$)	k (W m^{-1} $^\circ$K^{-1})	κ (m^2 s^{-1} $\times 10^4$)	Pr
Oxygen							
100	3.9918	0.9479	7.768×10^{-6}	1.946	0.00903	0.023876	0.815
150	2.6190	0.9178	11.490	4.387	0.01367	0.05688	0.773
200	1.9559	0.9131	14.850	7.593	0.01824	0.10214	0.745
250	1.5618	0.9157	17.87	11.45	0.02259	0.15794	0.725
300	1.3007	0.9203	20.63	15.86	0.02676	0.22353	0.709
350	1.1133	0.9291	23.16	20.80	0.03070	0.2968	0.702
400	0.9755	0.9420	25.54	26.18	0.03461	0.3768	0.695
450	0.8682	0.9567	27.77	31.99	0.03828	0.4609	0.694
500	0.7801	0.9722	29.91	38.34	0.04173	0.5502	0.697
550	0.7096	0.9881	31.97	45.05	0.04517	0.6441	0.700
600	0.6504	1.0044	33.92	52.15	0.04832	0.7399	0.704
Nitrogen							
100	3.4808	1.0722	6.862×10^{-6}	1.971	0.009450	0.025319	0.786
200	1.7108	1.0429	12.947	7.568	0.01824	0.10224	0.747
300	1.1421	1.0408	17.84	15.63	0.02620	0.22044	0.713
400	0.8538	1.0459	21.98	25.74	0.03335	0.3734	0.691
500	0.6824	1.0555	25.70	37.66	0.03984	0.5530	0.684
600	0.5687	1.0756	29.11	51.19	0.04580	0.7486	0.686
700	0.4934	1.0969	32.13	65.13	0.05123	0.9466	0.691
800	0.4277	1.1225	34.84	81.46	0.05609	1.1685	0.700
900	0.3796	1.1464	37.49	91.06	0.06070	1.3946	0.711
1000	0.3412	1.1677	40.00	117.2	0.06475	1.6250	0.724
1100	0.3108	1.1857	42.28	136.0	0.06850	1.8591	0.736
1200	0.2851	1.2037	44.50	156.1	0.07184	2.0932	0.748

Table B.2. Physical properties of saturated liquids

t (°C)	ϱ (kg m^{-3})	c_p (kJ kg^{-1} °K^{-1})	ν (m^2 s^{-1})	k (W m^{-1} °K^{-1})	κ (m^2 s^{-1} ×10^7)	Pr	β (°K^{-1})
Ammonia, NH$_3$							
−50	703.69	4.463	0.435×10^{-6}	0.547	1.742	2.60	
−40	691.68	4.467	0.406	0.547	1.775	2.28	
−30	679.34	4.476	0.387	0.549	1.801	2.15	
−20	666.69	4.509	0.381	0.547	1.819	2.09	
−10	653.55	4.564	0.378	0.543	1.825	2.07	
0	640.10	4.635	0.373	0.540	1.819	2.05	
10	626.16	4.714	0.368	0.531	1.801	2.04	
20	611.75	4.798	0.359	0.521	1.775	2.02	2.45×10^{-3}
30	596.37	4.890	0.349	0.507	1.742	2.01	
40	580.99	4.999	0.340	0.493	1.701	2.00	
50	564.33	5.116	0.330	0.476	1.654	1.99	
Carbon dioxide, CO$_2$							
−50	1156.34	1.84	0.119×10^{-6}	0.0855	0.4021	2.96	
−40	1117.77	1.88	0.118	0.1011	0.4810	2.46	
−30	1076.76	1.97	0.117	0.1116	0.5272	2.22	
−20	1032.39	2.05	0.115	0.1151	0.5445	2.12	
−10	983.38	2.18	0.113	0.1099	0.5133	2.20	
0	926.99	2.47	0.108	0.1045	0.4578	2.38	
10	860.03	3.14	0.101	0.0971	0.3608	2.80	
20	772.57	5.0	0.091	0.0872	0.2219	4.10	14.00×10^{-3}
30	597.81	36.4	0.080	0.0703	0.0279	28.7	
Dichlorodifluoromethane (Freon), CCl$_2$F$_2$							
−50	1546.75	0.8750	0.310×10^{-6}	0.067	0.501	6.2	2.63×10^{-3}
−40	1518.71	0.8847	0.279	0.069	0.514	5.4	
−30	1489.56	0.8956	0.253	0.069	0.526	4.8	
−20	1460,57	0.9073	0.235	0.071	0.539	4.4	
−10	1429.49	0.9203	0.221	0.073	0.550	4.0	
0	1397.45	0.9345	0.214	0.073	0.557	3.8	
10	1364.30	0.9496	0.203	0.073	0.560	3.6	
20	1330.18	0.9659	0.198	0.073	0.560	3.5	
30	1295.10	0.9835	0.194	0.071	0.560	3.5	
40	1257.13	1.0019	0.191	0.069	0.555	3.5	
50	1215.96	1.0216	0.190	0.067	0.545	3.5	

Table B.2 (continued)

t (°C)	ϱ (kg m^{-3})	c_p (kJ kg^{-1} °K^{-1})	ν (m^2 s^{-1})	k (W m^{-1} °K^{-1})	κ (m^2 s^{-1} ×10^7)	Pr	β (°K^{-1})
Engine oil (unused)							
0	899.12	1.796	0.00428	0.147	0.911	47100	
20	888.23	1.880	0.00090	0.145	0.872	10400	0.70×10^{-3}
40	876.05	1.964	0.00024	0.144	0.834	2870	
60	864.04	2.047	0.839×10^{-4}	0.140	0.800	1050	
80	852.02	2.131	0.375	0.138	0.769	490	
100	840.01	2.219	0.203	0.137	0.738	276	
120	828.96	2.307	0.124	0.135	0.710	175	
140	816.94	2.395	0.080	0.133	0.686	116	
160	805.89	2.483	0.056	0.132	0.663	84	
Ethylene glycol, $C_2H_4(OH)_2$							
0	1130.75	2.294	57.53×10^{-6}	0.242	0.934	615	
20	1116.65	2.382	19.18	0.249	0.939	204	0.65×10^{-3}
40	1101.43	2.474	8.69	0.256	0.939	93	
60	1087.66	2.562	4.75	0.260	0.932	51	
80	1077.56	2.650	2.98	0.261	0.921	32.4	
100	1058.50	2.742	2.03	0.263	0.908	22.4	
Eutectic calcium chloride solution, 29.9% $CaCl_2$							
−50	1319.76	2.608	36.35×10^{-6}	0.402	1.166	312	
−40	1314.96	2.6356	24.97	0.415	1.200	208	
−30	1310.15	2.6611	17.18	0.429	1.234	139	
−20	1305.51	2.688	11.04	0.445	1.267	87.1	
−10	1300.70	2.713	6.96	0.459	1.300	53.6	
0	1296.06	2.738	4.39	0.472	1.332	33.0	
10	1291.41	2.763	3.35	0.485	1.363	24.6	
20	1286.61	2.788	2.72	0.498	1.394	19.6	
30	1281.96	2.814	2.27	0.511	1.419	16.0	
40	1277.16	2.839	1.92	0.523	1.445	13.3	
50	1272.51	2.868	1.65	0.535	1.468	11.3	
Glycerin, $C_3H_6(OH)_3$							
0	1276.03	2.261	0.00831	0.282	0.983	84.7×10^3	
10	1270.11	2.319	0.00300	0.284	0.965	31.0	
20	1264.02	2.386	0.00118	0.286	0.947	12.5	0.50×10^{-3}
30	1258.09	2.445	0.00050	0.286	0.929	5.38	
40	1252.01	2.512	0.00022	0.286	0.914	2.45	
50	1244.96	2.583	0.00015	0.287	0.893	1.63	

Table B.2 (continued)

t (°C)	ϱ (kg m^{-3})	c_p (kJ kg^{-1} °K^{-1})	ν (m^2 s^{-1})	k (W m^{-1} °K^{-1})	κ (m^2 s^{-1} ×10^7)	Pr	β (°K^{-1})
Mercury, Hg							
0	13,628.22	0.1403	0.124×10^{-6}	8.20	42.99	0.0288	
20	13,579.04	0.1394	0.114	8.69	46.06	0.0249	1.82×10^{-4}
50	13,505.84	0.1386	0.104	9.40	50.22	0.0207	
100	13,384.58	0.1373	0.0928	10.51	57.16	0.0162	
150	13,264.28	0.1365	0.0853	11.49	63.54	0.0134	
200	13,144.94	0.1570	0.0802	12.34	69.08	0.0116	
250	13,025.60	0.1357	0.0765	13.07	74.06	0.0103	
315.5	12,847	0.134	0.0673	14.02	81.5	0.0083	
Methyl chloride, CH$_3$Cl							
−50	1052.58	1.4759	0.320×10^{-6}	0.215	1.388	2.31	
−40	1033.35	1.4826	0.318	0.209	1.368	2.32	
−30	1016.53	1.4922	0.314	0.202	1.337	2.35	
−20	999.39	1.5043	0.309	0.196	1.301	2.38	
−10	981.45	1.5194	0.306	0.187	1.257	2.43	
0	962.39	1.5378	0.302	0.178	1.213	2.49	
10	942.36	1.5600	0.297	0.171	1.166	2.55	
20	923.31	1.5860	0.293	0.163	1.112	2.63	
30	903.12	1.6161	0.288	0.154	1.058	2.72	
40	883.10	1.6504	0.281	0.144	0.996	2.83	
50	861.15	1.6890	0.274	0.133	0.921	2.97	
Sulfur dioxide, SO$_2$							
−50	1,560.84	1.3595	0.484×10^{-6}	0.242	1.141	4.24	
−40	1,536.81	1.3607	0.424	0.235	1.130	3.74	
−30	1,520.64	1.3616	0.371	0.230	1.117	3.31	
−20	1,488.60	1.3624	0.324	0.225	1.107	2.93	
−10	1,463.61	1.3628	0.288	0.218	1.097	2.62	
0	1,438.46	1.3636	0.257	0.211	1.081	2.38	
10	1,412.51	1.3645	0.232	0.204	1.066	2.18	
20	1,386.40	1.3653	0.210	0.199	1.050	2.00	1.94×10^{-3}
30	1,359.33	1.3662	0.190	0.192	1.035	1.83	
40	1,329.22	1.3674	0.173	0.185	1.019	1.70	
50	1299.10	1.3683	0.162	0.177	0.999	1.61	

Table B.2 (continued)

t (°C)	ϱ (kg m^{-3})	c_p (kJ kg^{-1} °K^{-1})	ν (m^2 s^{-1})	k (W m^{-1} °K^{-1})	κ (m^2 s^{-1} ×10^7)	Pr	β (°K^{-1})
Water, H$_2$O							
0	1002.28	4.2178	1.788×10^{-6}	0.552	1.308	13.6	
20	1000.52	4.1818	1.006	0.597	1.430	7.02	0.18×10^{-3}
40	994.59	4.1784	0.658	0.628	1.512	4.34	
60	985.46	4.1843	0.478	0.651	1.554	3.02	
80	974.08	4.1964	0.364	0.668	1.636	2.22	
100	960.63	4.2161	0.294	0.680	1.680	1.74	
120	945.25	4.250	0.247	0.685	1.708	1.446	
140	928.27	4.283	0.214	0.684	1.724	1.241	
160	909.69	4.342	0.190	0.680	1.729	1.099	
180	889.03	4.417	0.173	0.675	1.724	1.004	
200	866.76	4.505	0.160	0.665	1.706	0.937	
220	842.41	4.610	0.150	0.652	1.680	0.891	
240	815.66	4.756	0.143	0.635	1.639	0.871	
260	785.87	4.949	0.137	0.611	1.577	0.874	
280.6	752.55	5.208	0.135	0.580	1.481	0.910	
300	714.26	5.728	0.135	0.540	1.324	1.019	

Table B.3. Physical properties of liquid metals

Metal	Melting point °C	Boiling point °C	T °C	ϱ kg m^{-3}	c_p kJ kg^{-1} °C^{-1}	$\mu \times 10^4$ kg m^{-1} s^{-1}	$\nu \times 10^6$ m^2 s^{-1}	k W m^{-1} °C^{-1}	$\kappa \times 10^6$ m^2 s^{-1}	Pr
Bismuth	271	1477	315	10011	0.144	16.2	0.160	16.4	11.25	0.0142
			538	9739	0.155	11.0	0.113	15.6	10.34	0.0110
			760	9467	0.165	7.9	0.083	15.6	9.98	0.0083
Lead	327	1737	371	10540	0.159	2.40	0.023	16.1	9.61	0.024
			704	10140	0.155	1.37	0.014	14.9	9.48	0.0143
Lithium	179	1317	204.4	509.2	4.365	5.416	1.1098	46.37	20.96	0.051
			315.6	498.8	4.270	4.465	0.8982	43.08	20.32	0.0443
			426.7	489.1	4.211	3.927	0.8053	38.24	18.65	0.0432
			537.8	476.3	4.171	3.473	0.7304	30.45	15.40	0.0476
Mercury	−38.9	357	−17.8	13707.1	0.1415	18.334	0.1342	9.76	5.038	0.0266
			93.3	13409.4	0.1365	12.224	0.0903	10.38	5.619	0.0161
			204.4	13168.1	0.1356	10.046	0.0748	12.63	7.087	0.0108
Sodium	97.8	883	93.3	931.6	1.384	7.131	0.7689	84.96	56.29	0.0116
			204.4	907.5	1.339	4.521	0.5010	80.81	66.80	0.0075
			315.6	878.5	1.304	3.294	0.3766	75.78	66.47	0.00567
			426.7	852.8	1.277	2.522	0.2968	69.39	64.05	0.00464
			537.8	823.8	1.264	2.315	0.2821	64.37	62.09	0.00455
			648.9	790.0	1.261	1.964	0.2496	60.56	61.10	0.00408
			760.0	767.5	1.270	1.716	0.2245	56.58	58.34	0.00385
Potassium	63.9	760	426.7	741.7	0.766	2.108	0.2839	39.45	69.74	0.0041
			537.8	714.4	0.762	1.711	0.2400	36.51	67.39	0.0036
			648.9	690.3	0.766	1.463	0.2116	33.74	64.10	0.0033
			760.0	667.7	0.783	1.331	0.1987	31.15	59.86	0.0033
NaK (56% Na, 44% K)	−11.1	784	93.3	889.8	1.130	5.622	0.6347	25.78	25.76	0.0246
			204.4	865.6	1.089	3.803	0.4414	26.47	28.23	0.0155
			315.6	838.3	1.068	2.935	0.3515	27.17	30.50	0.0115
			426.7	814.2	1.051	2.150	0.2652	27.68	32.52	0.0081
			537.8	788.4	1.047	2.026	0.2581	27.68	33.71	0.0076
			648.9	759.5	1.051	1.695	0.2240	27.68	34.86	0.0064

Table B.4. Physical Properties of Metals

Metal	Melting point °C	Properties at 20°C				Thermal conductivity k (W m^{-1} °C^{-1})								
		ϱ kg m^{-3}	c_p kJ kg^{-1} °C^{-1}	k W m^{-1} °C^{-1}	κ m^2 s^{-1} ×10^5	−100°C	0°C	100°C	200°C	300°C	400°C	600°C	800°C	1000°C
Aluminium:														
Pure	660	2707	0.896	204	8.418	215	202	206	215	228	249			
Al-Cu (Duralumin) 94–96% Al, 3–5% Cu, trace Mg		2787	0.883	164	6.676	126	159	182	194					
Al-Si (Silumin, copper-bearing), 86.5% Al, 1% Cu		2659	0.867	137	5.933	119	137	144	152	161				
Al-Si (Alusil), 78–80% Al, 20–22% Si		2627	0.854	161	7.172	144	157	168	175	178				
Al-Mg-Si, 97% Al, 1% Mg, 1% Si, 1% Mn		2707	0.892	177	7.311		175	189	204					
Beryllium	1277	1850	1.825	200	5.92									
Bismuth	272	9780	0.122	7.86	0.66									
Cadmium	321	8650	0.231	96.8	4.84									

Table B.4 (continued)

Metal	Melting point °C	Properties at 20°C				Thermal conductivity k (W m^{-1} °C^{-1})									
		ϱ kg m^{-3}	c_p kJ kg^{-1} °C^{-1}	k W m^{-1} °C^{-1}	κ m^2 s^{-1} ×10^5	−100°C	0°C	100°C	200°C	300°C	400°C	600°C	800°C	1000°C
Copper:														
Pure	1085	8954	0.3831	386	11.234	407	386	379	374	369	363	353		
Aluminium bronze 95% Cu, 5% Al		8666	0.410	83	2.330									
Bronze 75% Cu, 25% Sn		8666	0.343	26	0.859									
Red brass 85% Cu, 9% Sn, 6% Zn		8714	0.385	61	1.804		59	71						
Brass 70% Cu, 30% Zn		8522	0.385	111	3.412	88		128	144	147	147			
German silver 62% Cu, 15% Ni, 22% Zn		8618	0.394	24.9	0.733	19.2		31	40	45	48			
Constantan 60% Cu, 40% Ni		8922	0.410	22.7	0.612	21		22.2	26					
Iron:														
Pure	1537	7897	0.452	73	2.034	87	73	67	62	55	48	40	36	35
Wrought iron, 0.5% C		7849	0.46	59	1.626		59	57	52	48	45	36	33	33
Steel (C max≈1.5%):														
Carbon steel														
C≈0.5%		7833	0.465	54	1.474		55	52	48	45	42	35	31	29
1.0%		7801	0.473	43	1.172		43	43	42	40	36	33	29	28
1.5%		7753	0.486	36	0.970		36	36	36	35	33	31	28	28

Table B.4 (continued)

Metal	Melting point °C	Properties at 20 °C				Thermal conductivity k (W m^{-1} °C^{-1})								
		ϱ kg m^{-3}	c_p kJ kg^{-1} °C^{-1}	k W m^{-1} °C^{-1}	κ m^2 s^{-1} ×10^5	−100 °C	0 °C	100 °C	200 °C	300 °C	400 °C	600 °C	800 °C	1000 °C
Nickel steel														
Ni≈ 0%		7897	0.452	73	2.026									
20%		7933	0.46	19	0.526									
40%		8169	0.46	10	0.279									
80%		8618	0.46	35	0.872									
Invar 36% Ni		8137	0.46	10.7	0.286									
Chrome steel														
Cr≈ 0%		7897	0.452	73	2.026	87	73	67	62	55	48	40	36	35
1%		7865	0.46	61	1.665		62	55	52	47	42	36	33	33
5%		7833	0.46	40	1.110		40	38	36	36	33	29	29	29
20%		7689	0.46	22	0.635		22	22	22	22	24	24	26	29
Cr-Ni (chrome-nickel): 15% Cr, 10% Ni		7865	0.46	19	0.527									
18% Cr, 8% Ni (V2A)		7817	0.46	16.3	0.444		16.3	17	17	19	19	22	27	31
20% Cr, 15% Ni		7833	0.46	15.1	0.415									
25% Cr, 20% Ni		7865	0.46	12.8	0.361									
Tungsten														
W = 0%		7897	0.452	73	2.026									
1%		7913	0.448	66	1.858									
5%		8073	0.435	54	1.525									
10%		8314	0.419	48	1.391									
Lead	328	11373	0.130	35	2.343	36.9	35.1	33.4	31.5	29.8				

Table B.4 (continued)

Metal	Melting point °C	Properties at 20 °C ρ kg m⁻³	c_p kJ kg⁻¹ °C⁻¹	k W m⁻¹ °C⁻¹	κ m² s⁻¹ ×10⁵	Thermal conductivity k (W m⁻¹ °C⁻¹) −100°C	0°C	100°C	200°C	300°C	400°C	600°C	800°C	1000°C
Magnesium:														
Pure	650	1746	1.013	171	9.708	178	171	168	163	157				
Mg-Al (electrolytic) 6–8% Al, 1–2% Zn		1810	1.00	66	3.605									
Molybdenum	2621	10220	0.251	123	4.790	138	125	118	114	111	109	106	102	99
Nickel:														
Pure (99.9%)	1455	8906	0.4459	90	2.266	104	93	83	73	64	59			
Ni-Cr 90% Ni, 10% Cr		8666	0.444	17	0.444		17.1	18.9	20.9	22.8	24.6			
80% Ni, 20% Cr		8314	0.444	12.6	0.343		12.3	13.8	15.6	17.1	18.0	22.5		
Silver:														
Purest	962	10524	0.2340	419	17.004	419	417	415	412		360			
Pure (99.9%)		10525	0.2340	407	16.563	419	410	415	374	362				
Tin, pure	232	7304	0.2265	64	3.884	74	65.9	59	57					
Tungsten	3387	19350	0.1344	163	6.271		166	151	142	133	126	112	76	
Uranium	1133	19070	0.116	27.6	1.25									
Zinc, pure	420	7144	0.3843	112.2	4.106	114	112	109	106	100	93			

C. Gamma, Beta and Incomplete Beta Functions

Gamma function definition

$$\Gamma(\alpha) = \int_0^\infty t^{\alpha-1} e^t dt$$

Recursion formula:

$$\Gamma(\alpha+1) = \alpha \Gamma(\alpha)$$

α	$\Gamma(\alpha)$	α	$\Gamma(\alpha)$	α	$\Gamma(\alpha)$
1.00	1.0000	1.35	0.8912	1.70	0.9086
1.05	0.9735	1.40	0.8873	1.75	0.9191
1.10	0.9514	1.45	0.8857	1.80	0.9314
1.15	0.9330	1.50	0.8862	1.85	0.9456
1.20	0.9182	1.55	0.8889	1.90	0.9618
1.25	0.9064	1.60	0.8935	1.95	0.9799
1.30	0.8975	1.65	0.9001	2.00	1.0000

Beta function definition:

$$B_1(\alpha, \beta) = \int_0^1 t^{\alpha-1}(1-t)^{\beta-1} dt = \frac{\Gamma(\alpha)\Gamma(\beta)}{\Gamma(\alpha+\beta)} = B_1(\beta, \alpha)$$

Incomplete Beta function definition:

$$B_x(\alpha, \beta) = \int_0^x t^{\alpha-1}(1-t)^{\beta-1} dt$$

Recursion formula:

$$B_x(\alpha, \beta) = B_1(\alpha, \beta) - B_{1-x}(\alpha, \beta)$$

The following table [1] gives the functional ratios $I_x(\alpha, \beta) = B_x(\alpha, \beta)/B_1(\alpha, \beta)$ for typical combinations of α and β:

Incomplete beta function ratios $I_x(\alpha, \beta)$

x	$\alpha = 1/3$ $\beta = 2/3$	$\alpha = 1/3$ $\beta = 4/3$	$\alpha = 1/3$ $\beta = 8/3$	$\alpha = 2/3$ $\beta = 4/3$	$\alpha = 1/9$ $\beta = 8/9$	$\alpha = 1/9$ $\beta = 10/9$	$\alpha = 1/9$ $\beta = 20/9$	$\alpha = 8/9$ $\beta = 10/9$
0	0	0	0	0	0	0	0	0
0.02	0.2249	0.3068	0.4007	0.0912	0.6346	0.6588	0.7281	0.0342
0.04	0.2838	0.3859	0.5007	0.1443	0.6856	0.7113	0.7845	0.0628
0.06	0.3254	04410	0.5684	0.1886	0.7173	0.7439	0.8186	0.0917
0.08	0.3588	0.4845	0.6204	0.2278	0.7407	0.7679	0.8431	0.1174
0.10	0.3872	0.5210	0.6627	0.2636	0.7595	0.7870	0.8622	0.1416
0.20	0.4924	0.6506	0.8008	0.4124	0.8213	0.8490	0.9199	0.2607
0.30	0.5694	0.7377	0.8793	0.5321	0.8603	0.8870	0.9506	0.3715
0.40	0.6337	0.8038	0.9284	0.6339	0.8895	0.9146	0.9696	0.4765
0.50	0.6911	0.8566	0.9599	0.7225	0.9133	0.9362	0.9820	0.5767
0.60	0.7448	0.8998	0.9796	0.7999	0.9335	0.9538	0.9901	0.6725
0.70	0.7970	0.9352	0.9912	0.8671	0.9515	0.9686	0.9952	0.7640
0.80	0.8501	0.9640	0.9972	0.9244	0.9679	0.9812	0.9982	0.8507
0.90	0.9084	0.9863	0.9996	0.9706	0.9835	0.9917	0.9996	0.9313
1.00	1.0000	1.0000	1.0000	1.0000	1.0000	1.0000	1.0000	1.0000
$B_1(\alpha,\beta)$	3.6275	2.6499	2.0153	1.2092	9.1853	8.8439	7.9839	1.0206

Reference

[1] Baxter, D. C, and Reynolds, W. C.: Fundamental solutions for heat transfer from non-isothermal flat plates. *J. Aero. Sci.* **25**:403, 1958.

The manufacturer's authorised representative in the EU is Springer Nature Customer Service Centre GmbH, Europaplatz 3, 69115 Heidelberg, Germany. If you have any concerns regarding our products, please contact ProductSafety@springernature.com

Printed and bound by CPI Group (UK) Ltd, Croydon, CR0 4YY

25/03/2026

02078189-0019